The Turing Test Argument

This book departs from existing accounts of Alan Turing's imitation game and test by placing Turing's proposal in its historical, social, and cultural context.

It reconstructs a controversy in England, 1946–1952, over the intellectual capabilities of digital computers, which led Turing to propose his test. It argues that the Turing test is best understood not as a practical experiment, but as a thought experiment in the modern scientific tradition of Galileo Galilei. The logic of the Turing test argument is reconstructed from the rhetoric of Turing's irony and wit. Turing believed that learning machines should be understood as a new kind of species, and their thinking should be understood as different from human thinking and yet capable of imitating it. He thought that the possibilities of the machines he envisioned were not utopian dreams. And yet he hoped that they would rival and surpass chauvinists and intellectuals who sacrifice independent thinking to maintain their power. These would be transformed into ordinary people, as work once considered 'intellectual' would be transformed into non-intellectual, 'mechanical' work.

The Turing Test Argument will appeal to scholars and students in the sciences and humanities and all those interested in Turing's vision of the future of intelligent machines in society and nature.

Bernardo Gonçalves is a visiting fellow at King's College, University of Cambridge, and a postdoctoral researcher at the University of São Paulo. His research focuses on Alan Turing and the future of machines in society and nature. He holds Ph.D. degrees in Philosophy and in Computational Modeling.

Routledge Studies in Twentieth-Century Philosophy

For more information about this series, please visit: www.routledge.com/
Routledge-Studies-in-Twentieth-Century-Philosophy/book-series/SE0438

The Turing Test Argument

Bernardo Gonçalves

Routledge
Taylor & Francis Group

NEW YORK AND LONDON

First published 2024
by Routledge
605 Third Avenue, New York, NY 10158

and by Routledge
4 Park Square, Milton Park, Abingdon, Oxon, OX14 4RN

Routledge is an imprint of the Taylor & Francis Group, an informa business

ISBN: 978-1-032-29157-4 (hbk)
ISBN: 978-1-032-29158-1 (pbk)
ISBN: 978-1-003-30026-7 (ebk)

DOI: 10.4324/9781003300267

Typeset in Sabon
by codeMantra

To Prof. Fabio Cozman, the vital source of support behind this work.

Do not let go of the guiding hand of history: history has made all, and history can change all.

—Ernst Mach, 1872

Contents

Acknowledgments

This book is a result of a very difficult career move that I decided to make when I left a permanent research position at IBM Research Brazil, where I was a computer scientist and artificial intelligence researcher, first on sabbatical leave and then permanently, to pursue my Turing research. I did this first in the form of a new Ph.D., in Philosophy, at the School of Philosophy, Languages, Literature and Humanities of the University of São Paulo, and then as a postdoctoral researcher at the Polytechnic School of the same university.

I owe my formative years in the humanities to many colleagues and friends, especially those I met at the University of São Paulo and the *Scientiae Studia* Association, which was an important center of learning for me from 2017 to 2019. I would especially like to acknowledge Marcos Camolezi, João Cortese, Orlando Pimentel, Marcos Barbosa, Scheilla Nunes, Marildo Menegat, Lucas Petroni, Dora Kaufman, Alexandre Moreli, Osvaldo Pessoa Jr., Pedro Bravo, Edelcio de Souza, Prof. Ricardo Terra, and Prof. Pablo Mariconda.

I am deeply grateful to Prof. Fabio Cozman, who has been an academic sponsor of my research since 2019. This book would not have been possible without his generous support and without the support of two postdoctoral research grants from the São Paulo Research Foundation — FAPESP grants 2022/16793-9 and 2019/21489-4, for the project 'The future of artificial intelligence: the logical structure of Alan Turing's argument' — which allowed me to dedicate years of continuous research to this project. I would also like to thank the FAPESP Area Coordinators who processed my research proposals and the anonymous reviewers who evaluated them; the people of the State of São Paulo, Brazil, whose hard work generates the public resources of the funding agency; and the Polytechnic School of the University of São Paulo. I would also like to express my deep gratitude to Prof. Murray Shanahan of Imperial College and Prof. Robin Osborne, the Vice-Provost of King's College, University of Cambridge.

For their kind assistance with archival access and permissions, I would like to thank Tom Davies and Patricia McGuire of King's College, Cambridge, and David Haberstich and Stephanie Kurasz of the National Museum of American History. The credit lines for the used materials appear alongside them.

I would like to thank Elliott Hodgkin for his high-quality copyediting and support.

My special thanks go to Andrew Weckenmann, who believed in my work early on, and to two Routledge reviewers, whose careful reading, suggestions, and comments on my book proposal took it to a new level.

I am also grateful to the editors of *Annals of Science, Philosophy & Technology, Minds & Machines*, and *AI & Society* who handled my article submissions and to several anonymous reviewers from these journals whose feedback, comments, and criticisms helped to improve this work significantly.

Also for feedback on my research, I am grateful to Pio Garcia, whose suggestions were essential to my Turing research from the beginning; Mark Priestley, who also kindly shared some important sources in the history of computing; Edgar Daylight and Prof. Erhard Schüttpelz, from whose discussions my research benefited greatly; Prof. Susan Sterrett, who contributed by far the most sensitive, attentive, and insightful remarks on Turing; and Prof. Richard Staley, who has been a very kind critic, colleague, and mentor.

Finally, I am very happy to acknowledge my wife, Carolina, for her support during years of intense research and writing that resulted in this book, which owes much to her love, patience, and companionship.

1 Introduction

This chapter introduces the book, starting with Alan Turing's central question, whether machines can think, and his particular approach to addressing it by means of his test, followed by an outline of the contributions of the book as they are organized in each chapter, a presentation of the materials used, the corpus of Turing sources and a periodization of them, and finally the methods used in the book.

1.1 Can machines think?

Suppose there were some mechanism — let us call it *think₂* — that could be built into a machine and make it talk freely and significantly better than parrots. The mechanism may be different in nature from the one embodied in us humans — call it *think₁* —, but it may be just enough to make the machine indistinguishable from a human in a teletyped conversation.

In the context of a public debate that took place in England, 1946–1952, Alan Turing (1912–1954, Figure 1.1) developed his imitation game and test to substantiate what is here called *the Turing test argument*, claiming through a sharp, subtle move that machines will think₂. Although these two ways of thinking are different in nature, they are similar in that both can exhibit *intelligence*. But if the concept of intelligence is subject to prejudice and is reduced to what humans and only humans can do, then machines, while operating on the basis of think₂, would have to show that they can imitate think₁ in order to be accepted and *pass* as intelligent.

Turing suggested that if a machine could imitate think₁, to the point of impersonating cultural stereotypes that are in principle alien to it, to the point of sustaining an interesting discussion of literary works such as those of Charles Dickens, then it would *not* be nonsense to include think₂ in the extension of what we simply call thinking. After all, the common use of words would *naturally* shift to accommodate the new reality of the talking machines. This would not be because Turing necessarily wanted it, but essentially because it was part of the future he foresaw early digital computers leading to.

To appreciate the depth of Turing's move, it may be helpful to consider an aphorism from the *Philosophical Investigations* (Wittgenstein, 1953):

Figure 1.1 Alan Turing (1912–1954). (*Source*: Photographs of Alan Turing, copyright The Provost and Scholars of King's College Cambridge 2023. Archives Centre, King's College, Cambridge. Reproduced with permission.)

When philosophers use a word — 'knowledge,' 'being,' 'object,' 'I,' 'proposition/sentence,' 'name' — and try to grasp the *essence* of the thing, one must always ask oneself: is the word ever actually used in this way in the language in which it is at home? —

What *we* do is to bring words back from their metaphysical to their everyday use.

(Wittgenstein, 1953, §116, his emphasis)

Turing's naturalistic approach,[1] like that of Ludwig Wittgenstein (1889–1951), has often been seen as a form of behaviorism, as if grounding philosophy in everyday life and common, social experience could be confused with scientific psychology and experimental measurements on pigeons and rats.

Turing saw no problem in extending common language to say that a machine 'thinks,' if this is consistent with having intelligent machines among us in our everyday life and experience. From an empirical point of view, this may be seen as a circular argument. And yet it is *not* a trivial one. The reason it is not trivial is that it is based on a counterfactual — a bet on a change in the *natural order* of our world resulting from the future impact of digital computers, which Turing foresaw.

The character of Turing's philosophical contribution, this book will suggest, is comparable to that of Galileo Galilei (1564–1642). Turing, like Galileo, was a master of thought experiments, and he sought to shift what Stephen Toulmin (1922–2009) called an 'ideal of natural order' (1961,

p. 36), which determines what seems natural, on the one hand, and what requires explanation, on the other hand. Such an ideal, Toulmin noted, is important because it sets the agenda of discussion. In the case of Galileo, according to Elizabeth Anscombe (1919–2001), this shift was once the subject of a remarkable exchange:

> He [Wittgenstein] once greeted me with the question: 'Why do people say that it was natural to think that the sun went round the earth rather than that the earth turned on its axis?' I replied: 'I suppose, because it looked as if the sun went round the earth.' 'Well,' he asked, 'what would it have looked like if it had *looked* as if the earth turned on its axis?' This question brought it out that I had hitherto given no relevant meaning to 'it looks as if' in 'it looks as if the sun goes round the earth.' My reply was to hold out my hands with the palms upward, and raise them from my knees in a circular sweep, at the same time leaning backwards and assuming a dizzy expression. 'Exactly!' he said.
>
> (Anscombe, 1959, p. 151, her emphasis)

Wittgenstein seems to have suggested that the reason why people used to think that the sun goes around the earth was not because it was *natural*. Rather, it was because it was part of a naturalized order, an image or picture of the world that was ingrained in common sense.

To the question of the ethics of building think$_2$ into machines and the consequent changes to the social order, Turing gave a satirical answer that will be explored in this book. He denied that he was guilty of a 'sort of Promethean irreverence,' using irony to return the burden of proof and expressing his suspicion of those who identified themselves as the humanists in the mind-machine debate. Turing also suggested that the actual existence of a machine-based think$_2$ should inform our interpretation of think$_1$, namely, that if language learning and use can be characterized as a mechanical process, this should inform and possibly constrain the meaning and significance we attach to think$_1$ as the hallmark of being human. It is true that, as argued in the early reception of Turing's 1950 paper, this can be seen as a form of Wittgensteinian therapeutic positivism. Turing's argument is well-suited to undermine humanity's high opinion of itself, its own intelligence, and its reserved place in nature, especially its domination of nature in the Anthropocene, all of which may play a role in the divisions that have been created within the human race itself.

More than seventy years later, the Turing test argument still seems to be largely received with a certain dizziness, either exaggerated by Whiggish excitement (the great foundation of artificial intelligence) or dismissed outright as philosophically naive. Meanwhile, as machine intelligence continues to evolve and reach new milestones, albeit in some ways away from Turing's original views, its impact on society increases, and the discussion around the Turing test returns.

This book aims to shed light on Turing's argument that the ideal of natural order by which we think about minds and machines would shift in the presence of more sophisticated machines. It is a relatively modest first step in the context of a larger intellectual project about Turing's construction of machine intelligence (think$_2$) and its relation to human intelligence (think$_1$), in its philosophical, historical, social, and cultural roots.

Preliminary words are given about the Turing test and the new perspective this book provides on it (Section 1.2). Then, the contributions of the book are outlined as they appear in each chapter (Section 1.3), and the materials (the Turing sources) and methods (for the history of philosophy) used are presented (Sections 1.4 and 1.5). Incidentally, I hope that this early presentation of materials and methods might help scientists from artificial intelligence and the cognitive sciences, who may be among the readers of this book, to appreciate its contributions and to take a fresh, new look at the Turing test.

1.2 What is the Turing test?

Turing opened his seminal paper (1950) by proposing to replace the question 'can machines think?,' which he deemed 'too meaningless to deserve discussion.' The new question, considered to have a 'more accurate form,' would be based on what Turing called the 'imitation game,' and later in the same text, his 'test.'[2] Details about the new question and the exact settings for its evaluation slipped through Turing's text in a sequence of variations that defies interpretation. According to different interpretations of the various versions of the test, the machine must be able to imitate stereotypes of a woman, a man, or a human, beside a true representative of the kind, to deceive a human interrogator about its true nature. The new question is whether the interrogator, at a distance and having no physical contact whatsoever, would be able to distinguish the machine from the genuine individual through a conversation game. If not, the machine must be considered intelligent.[3]

Turing's imitation game and test, like Turing himself, have long since become cultural icons. But what did Turing mean to propose with his test? Seventy years later, this question still defies a generally accepted answer. This book aims to provide a well-informed answer, combining an extensive study of Turing's primary and secondary sources with biographical details of him and his contemporaries, while also seeking a dialogue with the secondary literature.

'The Turing test is a joke, sort of,' said the North American computer scientist Marvin Minsky (1927–2016) in an interview (2013). It was 'about saying,' Minsky added, that 'a machine would be intelligent if it does things that an observer would say must be being done by a human.' He concluded: 'it was suggested by Alan Turing as one way to evaluate a machine but he had never intended it as *the way to decide* whether a machine was really intelligent' (emphasis added). Now, if Minsky was right in his characterization, is the Turing test a philosophically empty joke, or can it be a joke with philosophical depth? Can't a joke be a rhetorical device to make a controversial point?

For a short answer, this book will draw attention to Turing's skillful use of irony in his discussions of the possibility of machine intelligence. Turing's irony will be identified and characterized here as satire, or irony with a point. In this connection, one aspect of Turing's milieu is worth mentioning: after World War II, particularly from 1947 on, he developed a close friendship with two low-profile figures who belonged to the circle of Cambridge Apostles.[4] These were the colorful and tender-hearted mathematician who would become his doctoral student, Robin Gandy (1919–1995), and the penetrating biographer and literary critic, Nicholas Furbank (1920–2014). Turing would make them his literary and will executors, respectively.[5] For comparison, it is said that Wittgenstein was invited to join the Cambridge Apostles by some of its high-profile members, notably by Keynes, and Russell feared that he would not appreciate the group's lack of seriousness and style of humor (McGuinness, 1988, p. 118). And yet, even the serious Wittgenstein is remembered to have once said: 'a serious and good philosophical work could be written that would consist entirely of *jokes* (without [being] facetious)' (Malcolm, 1958, p. 28, Wittgenstein's emphasis). Now, Turing was too queer for the Cambridge Apostles, and according to Hodges (1983), he was never invited to join — i.e., was rejected by — the élite group. This can only make Turing's humor, on the fringes of the Cambridge Apostles, all the more remarkable and interesting. Another aspect of Turing's irreverence was his metaphilosophical approach to philosophical problems: e.g., his shift from the original question of 'can machines think?' to the question of how humans would blindly judge a machine that could play the imitation game well, and his return of the burden of proof to those who wished to draw a line between human intelligence and machine intelligence. In this connection, it is worth noting that Turing had been an undergraduate and then a research fellow in mathematics and logic at Cambridge in the 1930s, at a time when Wittgenstein was a dominant figure in Anglophone philosophy and was shaking the foundations of mathematics and logic.[6] This is an important aspect of the backdrop of Turing's own irreverent, metaphilosophical approach to philosophical problems.

The sustained answer this book will give to the question of 'what the Turing test is' is that it is indeed something of a joke, but far from an intellectually empty one. Its scientific and philosophical depth can be appreciated by reconstructing it as a thought experiment in the modern scientific tradition of Galileo. In fact, it should be noted that the Turing test was proposed in the context of a scientific and philosophical controversy over the meaning and significance of early digital computers, which became a controversy about minds and machines. It is typical of thought experiments that they are proposed in the context of a controversy. It is well known that Turing chose conversation as the intellectual task to be tackled in his famous test. But it is often less appreciated that, beyond the rhetoric of the imitation game, *language learning* is the fundamental intellectual task that Turing proposed for the machine to master in preparation for the game. Further, Turing chose conversation, this book will suggest, perhaps not so much because he saw the sustained use of language as particularly characteristic of intelligence, but mainly because he realized that it was the ability that humans themselves associated with intelligence, as explicitly suggested to him by some of his interlocutors as part of the controversy just mentioned.

The interpretation of the Turing test presented in this book pays close attention to the following anecdote by Robin Gandy, who, as mentioned, was one of Turing's closest friends in his postwar life. Gandy wrote:

> It [Turing's *Mind* paper] was intended not so much as a penetrating contribution to philosophy but as propaganda. Turing thought the time had come for philosophers and mathematicians and scientists to take seriously the fact that computers were not merely calculating engines but were capable of behaviour which must be accounted as intelligent; he sought to persuade people that this was so. He wrote this paper unlike his mathematical papers quickly and with enjoyment. I can remember him reading aloud to me some of the passages always with a smile, sometimes with a giggle.
>
> (Gandy, 1996, p. 125)

If Turing sought propaganda, he seems to have succeeded. To the extent that bibliometrics can be informative, at the time of this writing, Oxford University Press counts 5944 citations to Turing's 1950 paper 'Computing Machinery and Intelligence.'[7] If Turing did not seek to make a penetrating contribution to philosophy, he seems to have given it anyway, as this book can suggest.

It is also worth noting that Gandy refers to 'philosophers and mathematicians and scientists' to whom Turing would have wanted to address. Who were they? There were three specific interlocutors with whom Turing sought a direct dialogue in his test proposal, and to whom this book gives unprecedented visibility, as they well deserve, in connection with the Turing test. These thinkers were (in the order of their first critical engagement with Turing):

- Douglas Hartree (1897–1958), a Fellow of the Royal Society (FRS) since 1932 and in 1949 was Professor of Mathematical Physics at the University of Cambridge in 1949.
- Geoffrey Jefferson (1886–1961), an FRS since 1947 and was Professor of Neurosurgery at the University of Manchester in 1949.
- Michael Polanyi (1913–1976), an FRS since 1944 and then Professor of Social Studies at the University of Manchester.

This book will argue that Gandy's anecdote agrees with both independent historical evidence and key structural elements of Turing's 1950 text. Gandy helps to expose the lack of historical depth in the analyses of the Turing test in the literature up to 2020. His first-hand account of Turing's aim in proposing his test differs significantly from the common conception of the Turing test in philosophy as a kind of crucial experiment to determine the existence of an intelligent machine.

1.3 The contributions and the organization of this book

This book makes the following contributions to the understanding of the Turing test and to Turing scholarship more generally.

1.3.1 Reception history

Chapter 2 presents a brief history of the reception of the Turing test and its evolution in the secondary literature since the publication of his 1950 paper in *Mind*, 'Computing Machinery and Intelligence.' It benefits from and extends the commentary in Stuart Shieber's 2004 edited volume, *The Turing Test: Verbal Behavior as the Hallmark of Intelligence*, which opened the way to philosophical historiographies of the Turing test. It included Turing's primary sources and collected, presented, and commented on some of the influential texts in the philosophical reception of Turing's 1950 paper up to the 1990s. Compared to Shieber, Chapter 2 focuses more on historical development and context, identifying cycles of the secondary literature on the Turing test, including influential texts outside of academic philosophy, and conceptual elements that mark each cycle.

To mention supporting bibliography that is not covered in the chapter, Andrew Hodges' Turing biography (1983) is a foundational reference on Turing's life and works and a mandatory reference for all Turing research. Hodges relied upon the first Turing biography, which was written by Turing's mother Mrs. Sara Turing (1959). Her great work and collection of sources deserves more credit. Jack Copeland contributed a carefully edited collection of Turing's most important writings, *The Essential Turing* (2004), whose pagination is used in this book for most quotations from Turing. Other accounts of Turing's life and works include (Swinton, 2019), (Copeland, 2006, 2012a), (Lavington, 2012), two large recent compendiums (Cooper and van Leeuwen, 2013; Copeland et al., 2017), and various other anthologies (Millican and Clark, 1999; Teuscher, 2006; Moor, 2003; Floyd, 2017). These must be distinguished from yet another kind of material (Epstein et al., 2009; Warwick and Shah, 2016), which focuses on the Turing Test and its practical implementations in a way that is not concerned with Turing's original views.

1.3.2   *Imitation principle*

Turing's imitation game and test was widely received as a reductionist view of intelligence, as if he were suggesting that (human) intelligence is a directly measurable quality. More recently, an opposing interpretation is that he understood (human) intelligence as an 'emotional' (socially constructed) concept and intended to test the human interrogators and judges, not the machines. Chapter 3 argues that both the reductionist and constructionist views downplay Turing's theoretical construction of machine intelligence and confuse his principle of imitation with identity. Two sides of the same coin, both interpret Turing's concept of intelligence as identical to either an operational definition or a socially accepted convention. Either way, it is as if Turing rejected a distinction between the imitation of behavior considered intelligent by human experimenters, interrogators, and judges, and intelligence itself as built into an autonomous agent. The chapter argues that these views of Turing's concept of intelligence are no more than cartoons that strip it of its contextual complexity, and that Turing conceived machine intelligence (think$_2$) in analogy to human intelligence (think$_1$), not identity. The proposed interpretation of Turing's scientific philosophy is supported by conceptual chronologies of primary sources. A first chronology, 1945–1952, shows that Turing based his concept of machine intelligence in his working mathematical theory of machine learning, inspired by the human development of a child's brain in its social environment. A second chronology, 1948–1952, clarifies why and in what terms Turing sought for an empirical basis for machine intelligence, an approach that could neutralize confirmation bias against the machine. Turing's concept of imitation was technical and overarching in his scientific philosophy. The epistemological importance of imitation, and in particular the establishment of the imitation of human intelligence as the empirical basis for machine intelligence, is emphasized.

Turing stated his belief that a suitable computer program for imitating human intelligence would eventually be found, and he pursued his research to inform our understanding of both nature *and* society, rather than instrumentally or for practical purposes. While he did *not* aim to literally reconstruct the human mind, the chapter argues, he did aim to construct think$_2$ to 'simulate the behaviour of the human mind very closely,' and he was intrigued to see if there was any intellectual human behavior that would ultimately remain beyond the reach of machines. He aimed to explore, by analogy and to the limit, the extension of mechanism as constitutive of the human mind, i.e., how mechanical the mind can be.

1.3.3   Postwar controversy

Turing's much-discussed test is over seventy years old and still fairly controversial. His 1950 paper is often considered a complex and multilayered text, and key questions about it remain largely unanswered. Why did Turing argue for learning from experience as the best approach to achieving machine intelligence? Why did he spend several years working with chess playing as a task to illustrate and test for machine intelligence, only to trade it out for conversational question-answering in 1950? Why did he refer to gender imitation in a test for machine intelligence?

Chapter 4, which is based on the article 'Can machines think? The Controversy that Led to the Turing Test' (Gonçalves, 2022), addresses these questions by revealing social, historical, and epistemological roots of the Turing test. I will draw attention to a historical fact that has received little attention in the secondary literature thus far, namely, that Turing wrote his *Mind* paper (1950) in the context of a specific controversy about the meaning and significance of early digital computers in England, or about minds and machines, that would last until 1952.

This observation departs from earlier interpretations of the Turing test in philosophy and opens the way to understanding that Turing's argument is largely *dialectical* in nature—it is a critical response to a series of specific objections raised against Turing's views at the time. His argument, despite its general philosophical value, was set in a specific dialogue with certain thinkers, notably his contemporaries mentioned above—Hartree, Jefferson, and Polanyi—, and to a lesser extent other public commentators in postwar Britain. This book argues that Turing's test proposal can hardly be understood without appreciating the context of that debate, which can shed light on several aspects of his 1950 text that might otherwise appear puzzling. In particular, Turing did propose gender learning and imitation as one of his various imitation tests for machine intelligence, and this was an important element in responding to Jefferson's suggestion that gendered behavior is causally related to the physiology of sex hormones.

1.3.4   Thought experiment

The Turing test has been studied and run as a controlled experiment and found to be underspecified and poorly designed. But it has also been defended and still attracts interest as a test for true artificial intelligence (AI). Scientists and philosophers regret the test's current status, acknowledging that the situation is at odds with the intellectual standards of Turing's works.

Chapter 5, which is based on the article 'The Turing Test is a Thought Experiment' (Gonçalves, 2023c), refers to this as the Turing Test Dilemma, following the observation that the test has been under discussion for over seventy years and still is widely seen as either too bad or too good to be a valuable experiment for AI. The chapter presents an argument that solves the dilemma by reconstructing the Turing test as a thought experiment in the modern scientific tradition. It shows that Turing's exposition of his imitation game and test, far from being an underspecified, poorly designed experiment, follows a methodical case and control structure. The presentation of the test satisfies what Ernst Mach (1838–1916) called 'the basic method of thought experiments,' characterized by a continuous variation of experimental conditions. Specifically, the shifts in Turing's presentation are designed to refute the fixation of ontological kinds, which he understands as fluid in nature and unjustly fixed in culture.[8] For example, he makes rhetorical use of the commonsense truism that a man can possibly imitate the intellectual stereotypes associated with a woman, despite their physical differences, to illustrate his point that the logical is not determined by the physical.

The chapter also argues that Turing's uses of his test are consistent with what has been called in the philosophical literature on thought experiments the 'critical' and 'heuristic' uses of thought experiments and the association of thought experiments with conceptual change. On the one hand, Turing used the imitation game to respond critically to the challenges posed to him by his contemporaries. It is emphasized how he methodically varied the design of the imitation game to address those challenges. On the other hand, he used it heuristically to illustrate a phenomenon about the concept of intelligence and to propose a hypothesis about machine learning, which he understood as the theoretical basis of machine intelligence. This reconstruction of the Turing test provides a rapprochement to the conflicting views on its value in the literature.

1.3.5   Galilean resonances

In 1950, Turing called his iconic imitation game a 'test,' an 'experiment,' and the 'the only really satisfactory support' for his view that machines can think. Following his own rhetoric, the 'Turing test' has been widely received as a kind of crucial experiment to determine machine intelligence.

In 1951 and 1952, however, Turing showed a milder attitude towards what he called as his 'imitation tests.' In 1948, Turing had referred to the persuasive power of 'the actual production of machines' rather than that of a controlled experiment.

Observing this, Chapter 6, which is based on the article 'Galilean Resonances: The Role of Experiment in Turing's Construction of Machine Intelligence' (Gonçalves, 2023a), proposes to distinguish the logical structure from the rhetoric of Turing's argument. I emphasize that experiment played an important role in Turing's construction of machine intelligence. But it can be shown that the implicit idea of a crucial experiment with demonstrative purposes never appears in his major report 'Intelligent Machinery' (1948) and actually appears specifically in Turing's 1950 paper, which itself, as mentioned, grew out of a specific controversy in England. I argue that Turing's proposal of a crucial experiment may have been a concession to meet the standards of his interlocutors more than his own, while his construction of machine intelligence rather reveals a method of successive idealizations and exploratory experiments. I will draw a parallel with Galileo's construction of idealized fall in a void and the historiographical controversies over the role of experiment in Galilean science. I suggest that Turing, like Galileo, relied on certain kinds of experiment, but also on rhetoric and propaganda to inspire further research that could lead to convincing scientific and technological progress.

1.3.6 *Irony with a point*

Turing made strong statements about the future of machines in society. Chapter 7, which is based on the article 'Irony with a Point: Alan Turing and his Intelligent Machine Utopia' (Gonçalves, 2023b), asks how they can be interpreted to advance our understanding of Turing's philosophy.

The chapter highlights Turing's systematic use of ironic insights, making arguments through the formulation of surprising contrasts intended to unsettle his interlocutors' assumptions. I argue that Turing sought irony as a method of self-expression but his irony has been largely caricatured or minimized by historians, philosophers, scientists, and others. Turing is often portrayed as an irresponsible scientist or associated with childlike manners and polite humor. While these representations of Turing have been widely disseminated, another image suggested by one of his contemporaries, that of a nonconformist, utopian, and radically progressive thinker reminiscent of the English Romantic poet Percy B. Shelley, has remained largely underexplored. Following this image, I will reconstruct the argument underlying what Turing called (but denied being guilty of) his 'Promethean irreverence' (1947–1951) as a utopian satire directed against chauvinists of all kinds, especially intellectuals who might sacrifice independent thought to maintain their power. These, Turing hoped, would eventually be rivaled and surpassed by intelligent machines and transformed into ordinary people, as work once considered 'intellectual' would be transformed into non-intellectual, 'mechanical' work.

I study Turing's irony in its historical context and follow the internal logic of his arguments to their limit. I suggest that Turing genuinely believed that the possibilities of the machines he envisioned were not utopian dreams. And yet he conceived them from a utopian frame of mind, aspiring to a different society, as a means of confronting what he saw as the social and cultural prejudices of his time and place, which are still alive today. His ever-learning child machines, whose intelligence would grow out of their own individual experiences, would help distribute power.

Organization of the chapters

Although each of the chapters introduced above has a different focus, they all share a common interpretation of the Turing test. For example, the interpretation of Turing's 1950 proposal as a thought experiment is informed by the fact that his proposal grew out of the controversy about the meaning and significance of digital computers or the controversy about minds and machines. This means that there should be some partial overlap in the content of the chapters. Some overlap is also justified to ensure that each chapter is fairly self-contained. Thus, the reader who goes through the whole book in one reading may have to forgive some redundancies, e.g., the reintroduction of Turing's contemporaries in a next chapter or coming across a recurring quotation because it is then analyzed from a different angle.

1.4   The Turing sources: corpus and periodization

The philosophical analyses presented in this book are grounded in a study of Turing's chronological construction of his concept of machine intelligence. Turing's imitation game and test for machine intelligence is, of course, an element of his wider conceptual framework for machine intelligence.

The original manuscripts of most of the primary sources cited in this book, comprising texts either written by Turing or transcribed from his lectures, are held and cataloged as 'The Papers of Alan Mathison Turing' (reference AMT) at the Archives Centre of King's College, Cambridge.[9] Facsimile copies of those original manuscripts are available at The Turing Digital Archive.[10] There are also the Alan Turing special collections at the University of Manchester Library in two parts. The first, the 'Alan Turing Collection,' contains digital copies of some of Turing's publications from Manchester, including a copy of his 'Mark I Programming Manual,' and some of his notes and working drafts relating to his work on computing and morphogenesis in plants.[11] The second part of the collection, 'Turing Additional,' is based on a file of academic and personal letters dating from early 1949 until Turing's death in June 1954, during his period at the University of Manchester.[12] The website maintained by Andrew Hodges also contains valuable information on Turing sources and may also be of interest to Turing studies.[13] It is worth noting that some of the key Turing sources have not been available or widely visible until more recently. For example, Turing's major report 'Intelligent Machinery' (1948) was buried in the archives of the NPL in Teddington near London until the late 1960s, when it was finally released from state secrecy and published (Evans and Robertson, 1968). At the time Turing delivered it, his boss, Charles Darwin (1887–1962), the director of the NPL and a grandson of the evolutionary thinker, thought it 'not suitable for publication.' Three new primary Turing sources were published in the late 1990s (Copeland, 1999). Minute notes of an event in 1949 that was crucial for Turing's formulation of the imitation game have only been published by Turing's contemporary Wolfe Mays (2000), who had the notes in his possession.[14] Secondary sources, mostly anecdotes from Turing's contemporaries, will also be used and discussed, not uncritically, especially for contextualizing and understanding the positions of Turing's intellectual opponents in the mind-machine controversy.

The following periodization is intended to provide a conceptual frame of reference against which Turing's construction of machine intelligence can be understood:

- The *foundational years I* (1936–1939): the period from Turing's paper 'On Computable Numbers' (1936) submitted on May 28, 1936, to his doctoral thesis at Princeton, published as a paper (1938) submitted on May 31, 1938, and his return to King's College, Cambridge.
- The *experimental years* (1939–1949): the period from Turing's wartime residence at Bletchley Park and his attendance at Wittgenstein's lectures in Cambridge in the spring of 1939 to the delivery of his report 'Intelligent Machinery' (1948) to the National Physical Laboratory (NPL) and the completion of his first year at the University of Manchester.
- The *dialogical years* (1949–1952): the period from his interview with *The Times* on June 11, 1949, to his round table discussion broadcast by the BBC on January 14, 1952.[15]
- The *foundational years II* (1951–1954): the period from Turing's letter to J. Z. Young of February 8, 1951, in which he reports on his work on

morphogenesis, a mathematical theory of the growth of living things, until his sudden, unexpected death in June 1954.[16]

This book focuses on Turing's *postwar years*, 1945–1952, which overlaps Turing's experimental and dialogical years and the second cycle of his foundational years. Turing's construction of machine intelligence was based on his mathematical theory of computing and the heuristic techniques he developed to solve empirical information problems during the war.

Two intellectual transitions are worth highlighting at this point. The first is Turing's move at the end of the war to 'build a brain' (Hodges, 1983; Sykes, 1992, p. 290; 25–27'), combining ideas from the foundations of mathematics (the first cycle of his foundational years) with empirical problem-solving (his experimental years). For one key aspect of this postwar transition, it is worth noting that Turing recast his concept of a universal machine from a dialectical concept used to refute David Hilbert's program for the complete mechanization of mathematics (1936) into a positive concept used to lay the conceptual foundations for the science and technology of early digital computing.[17] His wartime experience can explain this move. The second transition to highlight is Turing's move from his experimental to his dialogical years in the Fall of 1949, his first year in Manchester. Understanding this transition is crucial to understanding of the Turing test. For one key aspect of this transition, it is worth noting that early on, from Turing's first thoughts on machine intelligence during the war, and through his NPL lecture and reports (1945; 1947; 1948) up to the opening of the Manchester seminars in October 1949, Turing still had the game of chess as his chosen intellectual task to illustrate and test for machine intelligence. From his 1950 paper on up to the BBC round table broadcast (1952) in January 1952, Turing referred to conversation as the most suitable intellectual task for a machine test. The transition to his dialogical years can explain this move, as Chapters 4 and 5 will argue.

1.5 The methods used in this book

For over seventy years, the Turing test has been widely discussed in various disciplines such as literary studies, especially science fiction studies, the cognitive sciences, computer science and artificial intelligence, and philosophy, especially the philosophy of mind. But it has been less so discussed from the perspectives of the history of philosophy and the history and philosophy of science. This book examines the Turing test from these latter perspectives, and this in itself has proven to be a remarkable source of originality.

I begin by discussing methodologies for the history of philosophy and then move on to methodologies for the history and philosophy of science.

1.5.1   *History of philosophy*

According to Martial Gueroult (1891–1976) in his *Descartes' Philosophy Interpreted According to the Order of Reasons* (Gueroult, 1952), inspired by the methodological principles of René Descartes (1596–1650), 'what is important and useful in the books of the superior intellects does not consist in such or such thoughts that one can extract from them; the precious fruit that they enclose ought to come from the whole body of the work' (p. xix). In this sense, the historian of philosophy should seek not 'unconnected thoughts' but 'the sequence and linkage of reasons' or, in short, *the order of reasons* that gives structure to the philosophical system under study.[18] At the micro level of a single exegetical unit within a philosophical corpus—an essay, a lecture, a treatise—, the task of interpretation is to reconstruct a sequence of interrelated logical steps that are supposed to constitute the argument.

To promptly illustrate the value of such a methodology, consider Turing's *Mind* paper, 'Computing Machinery and Intelligence' (1950), which is the main object of analysis in this book. It has been said to be accessible to a general readership.[19] And yet, on close reading, it has been said to be a complex, multilayered text (Genova, 1994), or one that is too ambiguous for interpretation (Hayes and Ford, 1995; McDermott, 2014). For example, an intriguing puzzle about Turing's 1950 text is why he considered the original question, whether machines can think, to be 'too meaningless to deserve discussion' and proposed to replace it with a test, an experiment, and yet spent most of his paper (Sections 6 and 7, which make up almost 70% of the paper) discussing that very question. Now, this is the kind of exegetical question that the Gueroultian methodology can help address. This book will apply it to present a close, original reading of Turing's 1950 paper later in Chapter 6. Essentially, the logical development of Turing's 1950 text can be interpreted in three main steps: the *proposal*, the *science*, and the *discussion*, the latter having both a critical and a propositional component. In light of this reading, the exegetical question just mentioned will prove to be more tractable than it may have seemed before. Essentially, this book will argue that instead of presenting a (crucial) experiment for machine intelligence, as is often suggested in the secondary literature (cf. Chapter 2), Turing used his proposal, and the science behind it, as a means of conducting a *philosophical* discussion of the original question, 'Can machines think?'

In Gueroult's methodology, only the text and its logical development are important. The philosophical text is taken out of history, or synchronically, as opposed to along its development in time, or diachronically. The synchronic approach was criticized in French philosophy by Ferdinand Alquié (1906–1985), who followed a diachronic approach and disputed with Gueroult the interpretation of Descartes' philosophy (see e.g. Peden, 2011). For Gueroult, the works of great philosophers, whether coming early or later in their philosophical careers, are expressions of their mature thought. In fact, Gueroult's methodology can be particularly limited for the study of Turing's philosophy, whose primary sources are hardly comparable to those of Descartes or, say, Kant or Hegel. Turing did not write polished treatises, nor did he leave us prolegomena. This is how Max Newman (1897–1984) described the fragmentary nature of Turing's works:

> The varied titles of Turing's published work disguise its unity of purpose. The central problem with which he started, and to which he constantly returned, is the extent and the limitations of mechanistic explanations of nature. All his work, except for three papers in pure mathematics … grew naturally out of the technical problems encountered in these inquiries.
>
> (Newman, 1955, p. 256)

It is remarkable that Newman, a close contemporary of Turing, could see a unity of purpose in Turing's works, despite their unsystematic nature.

Turing's writings can be compared to Gottfried Wilhelm Leibniz's writings in the terms described by Hector-Neri Castañeda (1974): 'a bewildering array of reflection … with all sorts of repetitions and tentative suggestions, often making a new start so as to capture a fragmentary insight' (p. 383). This differs significantly from Descartes' writings: complete, organized, systematic.

In dealing with the difficult exegetical issues in Leibniz's philosophy, Castañeda proposed a methodological distinction for the history of philosophy that implicitly echoes in Anglophone philosophy the disputes just mentioned in France. Castañeda described the 'Athenian conception' of the work of a great philosopher, which is reminiscent of Gueroult's and which deliberately excludes its chronological dimension from the analysis:

> Many studies in the history of philosophy assume, more or less explicitly, an *Athenian conception* of a great philosopher's work. By this I mean that somehow the idea is taken for granted that there is such a thing as *the* system of the great philosopher under study and that it, like Athena springing forth whole and adult from Zeus's head, was conceived by the philosopher more or less as one total unity from the very beginning.
>
> (Castañeda, 1974, p. 382, his emphasis)

The Athenian approach, he continued, is difficult and very demanding to practice rigorously, and it can easily be abused:

> Thus, the historian's task of reconstructing and laying bare the philosopher's system is made both difficult and easy: difficult because of the abundance of materials that become relevant for each claim and easy because the historian can pick and choose his passages, whatever their sizes, from *anywhere* in the philosopher's philosophical corpus in order to support his attributions of theses.
>
> (Castañeda, 1974, p. 382, his emphasis)

The Athenian approach, Castañeda found, 'is doubly distortive,' as 'it encourages the interpretation of short texts out of their contexts.' He went on to describe another approach to the history of philosophy, which he called 'Darwinian' to emphasize its evolutionary and diachronic perspective and to contrast it with the Athenian one. The philosophical corpus of a great philosopher, he wrote, 'is a complete *Darwinian fauna* (*ibid.*, his emphasis):

> [T]he Darwinian approach to a great philosopher's so-called philosophical system is both descriptive and constructive: it constructs systems out of the views, theses, and half-systems as they appear in the corpus and parades them in an evolutionary ranking, whatever the causes of development may have been.
>
> The views and systems or half-systems of each [exegetical] unit must be exhibited as fully as is possible and then compared with similar outcomes of the exegesis of other units. Thus, we *cannot* pick and choose any fragments whatever from the corpus in support of one another, unless they are evaluated and shown to represent the same point of view on a certain issue.
>
> (Castañeda, 1974, p. 382, his emphasis)

Further, the 'Darwinian' approach can benefit from the support of a conceptual chronology such as the one presented above to make sense of Turing's construction of machine intelligence over the years. For example, I will show that the traditional interpretation of Turing's test as a crucial experiment to determine the existence of an intelligent machine is in tension with the development of his thought over the years. It is arguably more defensible that he used a rhetoric of crucial experiment dialectically to meet the standards of his interlocutors more than his own (Chapter 6).

However, as mentioned above, the interpretation of the Turing Test argument presented in this book is also based on the structural reading proposed by Gueroult, which is applied explicitly to Turing's 1950 paper and implicitly to other exegetical units. For example, the sequence of reasons in Turing's 'Intelligent Machinery' (1948) is considered in order to dispute the interpretation of a key passage with Diane Proudfoot (2017a)—namely, Turing's statement that 'the idea of "intelligence" is itself emotional rather than mathematical' (1948, p. 411). Based on this passage, Proudfoot attributes to Turing the view that *no* mathematical theory of intelligence is possible.[20] In response, I argue that Turing could not have meant the interpretation that Proudfoot suggests for this passage, while at the same time spending most of his 1948 text articulating and proposing a mathematical and computational theory of intelligence.[21] In other words, the passage should not be arbitrarily selected and interpreted out of the exegetical unit and the broader context of Turing's chronological construction of machine intelligence.

For more discussion of conceptual and methodological problems in the history of philosophy, the reader may wish to consult, for example, (Beck, 1969; Rorty et al., 1984).

1.5.2   *History and philosophy of science*

A more detached view of the history of philosophy as a philosophical problem, almost as an anthropological problem, can be found in the philosophy of the later Wittgenstein, although it was not proposed in the same framework of discussion as that of Gueroult and others. Scholars such as Gerd Buchdahl (1914–2001) drew some inspiration from Wittgenstein's philosophy (Anderson, 2022).[22] Buchdahl led the formation of history and philosophy of science as a new discipline in Melbourne, Australia, and Cambridge, United Kingdom, emphasizing that 'a *critical* approach to the history of science will do well to avail itself of the results of philosophical scholarship; and on the other side, a study of philosophical concepts, particularly those appertaining to the field of science, must need see their development in the concrete contexts of historical reality' (Buchdahl, 1962, p. 64, his emphasis). This book follows Buchdahl's general approach to studies in the history and philosophy of science, which is used here in conjunction with the methods of the history of philosophy.

The analysis of the Turing test presented in this book is largely based on the study of the controversy in England, 1946–1952, over the meaning and significance of digital computers (from the point of view of science and the history of computing) or over minds and machines (from the point of view of analytic philosophy). That controversy has long been a neglected element in the discussions of the Turing test in the secondary literature, and this book is possibly the first attempt to address the Turing test from this angle. The controversy is used in this book as a method (Pinch, 2015, p. 282).

I strive for the preservation of *symmetry* in a sense that is inspired by, but not exactly the same as that of David Bloor's *Knowledge and Social Imagery* (1976) or that of Steven Shapin and Simon Schaffer's *Leviathan and the Air Pump* (1985). Trevor Pinch describes the symmetry principle in this way:

> This principle called upon sociologists to use the same explanatory resources to explain both successful and unsuccessful knowledge claims. This was a way to avoid a sociology of error whereby the social would be evoked to explain false knowledge, leaving true knowledge as simply revealed, or generated by some universal epistemological principle or rule of method. Controversies at the research frontiers over knowledge claims were good places to carry through the principle of 'symmetry.' With each side in a dispute claiming to have 'truth on its side,' and disparaging the efforts of the other, and with the very outcome of the controversy still unknown, there was less temptation to carry out a sociology of error.
>
> (Pinch, 2015)

An important aspect of the controversy examined in this book is that it took place not long ago, in the middle of the twentieth century, and it can hardly be said to be settled. Thus, the distinction between 'successful' and 'unsuccessful' knowledge claims does not clearly apply here when we consider contemporary narratives about the meaning and significance of digital computers or about the nature of mind and machine. But if there is one historical actor who could possibly be characterized as 'unsuccessful' in his claims *at the time*, it is Alan Turing. This will be reflected in the early reception of his test proposal (Chapter 2). It is important to note that the construal of Turing's portrait as 'the father' of computer science is postmortem and was mobilized to serve specific interests in the formation of computer science as a discipline—see (Daylight, 2014; Bullynck et al., 2015; Daylight, 2015). Likewise, although Turing certainly influenced pioneers in artificial intelligence as an emerging discipline in the US in the 1950s, Turing's 'fatherhood' can also be questioned here. For one point of tension, it can be argued that the North American pioneers received Turing's views on machine intelligence through the lens of a distorted view of the 'Turing machine'—for example, see (Daylight, 2023).[23]

If Turing's contributions have been distorted and he has been made a hero of Whiggish scientific progress in computer science, artificial intelligence, and artificial life, none of this should detract from Turing's image as a brilliant, bold, true mathematician, scientist, and philosopher, all of which he certainly was—for an interesting account of this contrast, see e.g. (Longo, 2018). In his own time and place, Turing's claims went against conventional wisdom, and he could not back them up with a practical, crucial experiment. Not only was Turing less prominent than his direct opponents, all of whom were Fellows of the Royal Society (FRS) and holders of university chairs by 1949–1950, when Turing proposed his test. Turing would become an FRS only later in the controversy, and with the help of Bertrand Russell, who did not hesitate to support eccentric intellectuals such as Turing and Wittgenstein. Chapter 6 will explain one particular move by Turing that, inadvertently or not, helped him to gain Russell's support.

Further, unlike in Shapin and Schaffer's major work (1985), the scientific instrument, in the case of this book the early digital computer, will be somewhat less central to the analysis. The reason is clear: the Turing test argument, as we will see, is not based on an actual experiment with a digital computer. Rather, the focus will be on how Turing conceived of a kind of experiment that only conceptually refers to digital computers, as a way of responding to his interlocutors in the controversy. Symmetry is sought here to avoid the historical and sociological error of denying Turing's interlocutors a voice or of assuming that there was a winner in their debates, which is clearly not true. As noted above, their disputes, while not devoid of broader philosophical and scientific implications, can be seen as reflecting different perspectives on postwar British society. In this book, their points of contention are seen to be as alive today as they were in the late 1940s and the early 1950s.

Another important difference with respect to Bloor's and Shapin and Schaffer's works is that the mind-machine controversy is not treated here as a means for a broader discussion of the sociology of science as a human enterprise, but only as an excellent window through which to analyze and understand Turing's test proposal. Also, this book seeks a compromise between the hybrid methodology for the history of philosophy described above, which focuses on the logical and chronological development of a particular thinker's philosophy, and this other methodology for the history and philosophy of science just described, which focuses on the controversies themselves. This releases the book from an orthodox application of symmetry, in particular, from spending as much effort on presenting the philosophy of each of Turing's interlocutors as on presenting Turing's own philosophy.

2 Reception History, 1950–2020

Turing's imitation game and test is over seventy years old and has survived several rounds of criticism and defense. This chapter gives a brief overview of the reception of Turing's test proposal along the decades since the publication of his 1950 paper, 'Computing Machinery and Intelligence.' It benefits from and extends Stuart Shieber's 2004 philosophical anthology, *The Turing Test: Verbal Behavior as the Hallmark of Intelligence*. Compared to Shieber, this chapter will focus more on historical development and context. The goal here is not to provide a complete survey, but rather to identify periods in the broader secondary literature and to explore the important conceptual elements that characterize each period: from the early reception of Turing's proposal in the 1950s, often as neo-behaviorism in philosophy and as a preliminary definition of machine intelligence in the emerging studies of artificial intelligence (AI) in the United States; to its use in the debates on mechanism and consciousness from the 1960s to the 1990s; and to its rejection by computer scientists and AI researchers and its defense by philosophers from the 1990s to the 2010s.

2.1   The 1950s: the early reception

On December 1, 1950, Wittgenstein replied to Norman Malcolm (1911–1990), who had asked for his opinion on Turing's 1950 paper:[24]

> You're quite right, a mathematician by name of Turing attended my lectures in '39 (they were *pretty* poor!) & it's probably the same man who wrote the article you mention. I haven't read it but I imagine it's *no* leg-pull.
> Affectionately
> Ludwig
> (Malcolm, 1958, pp. 129–130, Wittgenstein's emphasis)

Wittgenstein was referring to his 1939 lectures on the foundations of mathematics (Diamond, 1976), which Turing attended. Some of their discussions appear in the notes of Wittgenstein's students, and a core part is quoted in (Hodges, 1983, pp. 152–154). As Wittgenstein's letter to Malcolm shows, he did not read Turing's paper (and he would die less than a year and a half later), and he did not need to read the paper to imagine that it was not a

DOI: 10.4324/9781003300267-2

leg-pull. The exact character of Wittgenstein's quasi-reception of Turing's paper may seem unclear, and if Wittgenstein could not read and properly comment on Turing's paper, one might say, so much the worse for Wittgenstein's legacy, as far as the twenty-first century and beyond is concerned. In any case, it is not unlikely that Wittgenstein was making a positive comment about Turing's work. First, the letter may suggest that Wittgenstein barely knew Turing, but this is not the case. In fact, according to Hodges (1983), who seems to have interviewed their common colleague, Alister G. D. Watson (1908–1982), they met sometimes in the botanic gardens of Cambridge in the summer of 1937 to discuss problems in the foundations of logic and mathematics. Second, Wittgenstein writes 'they were *pretty* poor,' but 'they' in this passage probably refers to his lectures delivered *ca.* eleven years earlier, since Turing appears to have been the only mathematician in the audience of the lectures.[25] The downplaying of his own earlier thoughts and writings in favor of the developmental aspect of his philosophy was not atypical of Wittgenstein.[26]

Bertrand Russell did read Turing's paper. In a letter to Turing dated April 19, 1951, Bertrand Russell's close friend Rupert Crawshay-Williams (1908-1977), who wrote *Russell Remembered* (1970), informed Turing of Russell's reception of Turing's 1950 paper. Russell and him 'read it and discussed it together,' and he wrote: 'We liked not only (of course) the general approach (the assumptions underlying your argument) but also the particular method and the examples.'[27] The letter also congratulated Turing on his election a month earlier as a Fellow of the Royal Society, sponsored by Turing's academic supporter Max Newman and Russell himself. Clearly, by that time, Russell had become a source of support for Turing's philosophy.

But Russell's positive early reception was not the rule. Turing's 1950 paper was largely received as a variant of behaviorism. Following Turing's election to the Royal Society, one of the first letters of congratulation he received was from one of his main intellectual opponents, Geoffrey Jefferson:

> I am so glad; and I sincerely trust that all your valves are glowing with satisfaction, and signalling messages that seem to you to mean pleasure and pride! (but don't be deceived!).[28]

Playing with words, Jefferson seems to allude to behaviorism as if it were related to Turing's views. Jefferson's letter of congratulations came less than six months after Turing's 1950 paper appeared in *Mind*, edited by Gilbert Ryle (1900–1976), whose views were associated with behaviorism. Ryle had just published what would become one of his best-known books, *The Concept of Mind* (1949). He wrote: 'The general trend of this book will undoubtedly, and harmlessly, be stigmatised as "behaviourist" (p. 327).

Turing's paper has often been read as a behaviorist account of intelligence. This is what Jefferson seems to be alluding to in his letter. In Ryle's words (*ibid.*), behaviorism states that scientific psychology should be based on repeatable and publicly checkable experiments, while emphasizing that the reputed deliverances of consciousness and introspection are not publicly checkable. Jefferson's interpretation of Turing as a behaviorist can explain his association of Turing with a machine whose introspective feelings could be misleading.

In professional philosophy, Turing's contemporary at the Department of Philosophy, University of Manchester, Wolfe Mays (1912–2005) published 'Can Machines Think?' (1952), one of the first reactions to Turing's 1950 paper.[29] Right in his abstract, Mays writes:

> Mr. A. M. Turing was quoted in *The Times* about a year ago as saying it would be interesting to discover the degree of intellectual activity of which a machine was capable and to what extent it could think for itself. He has now pressed this suggestion further and given the results of his researches in an article called 'Computing Machines and Intelligence,' together with a brief account of a 'child-machine' which he has attempted to educate (*Mind*, October 1950). I intend to discuss this article in some detail and examine his claim that 'machines can think."
>
> (Mays, 1952)

It is not hard to see that Mays felt provoked by Turing. After promptly casting Turing's proposal as 'modern neo-behaviorism' (1952, p. 149), Mays continued:

> Even if it were possible to construct a machine whose behaviour was indistinguishable from that of a human being, and even if we accept the behaviourist criterion it might still be useful to distinguish between men produced by natural methods and artificial men (or robots) ... The underlying assumption of many modern theories of intelligence is that intelligence is a simple quality and directly measurable. There are good grounds, however, for believing otherwise, that it is heterogeneous in character and only measurable by indirect methods.
>
> (Mays, 1952, p. 151)

Mays' formulation is suggestive. Did Turing regard intelligence as a simple, directly measurable quality? Or would he agree with Mays that it could only be measured by indirect methods? It seems that the delay in publication of Mays' paper from October 1950 to April 1952 jeopardized the possibility of a direct response from Turing, for by 1952 Turing had shifted the focus of his work somewhat, from the nature and capabilities of digital computers to a mathematical theory of the growth of living things.[30]

Mays went on to cite Jefferson, his and Turing's contemporary at Manchester University, whom Mays wanted to support:[31]

> From what has already been said, it will be seen that the question 'Can machines think?' means something very different than it does for Professor Jefferson. For Jefferson, and I should say for most ordinary people, any definition would also include certain psychological characteristics. Turing and Jefferson are in fact speaking different languages; in the behaviouristic (or physical) language of Turing — words which only have an objective physical content appear (or should appear), electronic tubes, flip-flop circuits, programmes, etc. It is a deterministic machine language in the grand manner of Newtonian physics.
>
> (Mays, 1952, p. 152)

Mays' association of Turing with behaviorism is clear and reminiscent of Jefferson's own provocations of Turing — Jefferson referred to Turing's 'valves,' and Mays referred to 'electronic tubes.' Mays seems to understand that Turing conflates analogy with identity.

Jefferson, in his Lister Oration (1949a), had blamed the 'physicists and mathematicians' for using an analogy between machines and brains only to slip into identity.[32] Jefferson's famous words quoted by *The Times* on June 10, 1949,[33] and later quoted by Turing (1950, p. 445) and Mays (1952, p. 151), were: 'Not until a machine can write a sonnet or compose a concerto because of thoughts and emotions felt, and not by the chance fall of symbols, could we agree that machine *equals* brain' (Jefferson, 1949a, p. 1110, my emphasis). A common aspect of Mays' and Jefferson's attacks on Turing is that they both understood Turing's test proposal as a claim that machines can *think*$_1$, that is, that they can think in the same identical way that humans can.[34] It is worth contrasting this with the understanding of Russell's associate, Rupert-Crawshay Williams. In his letter quoted above, he also wrote:

> I myself am particularly interested in your comments on Geoffrey Jefferson, who I think managed to get the concepts of similarity and difference all wrong. But that's too much to go into.[35]

Ryle rejected Mays' paper, but less than a year after publishing Turing's paper, he published a short paper by the geneticist Leonard Pinsky, 'Do Machines Think About Machines Thinking?' (1951). To satirize 'Mr. Turing's game,' Pinsky proposed to replace Turing's question of whether machines could play the imitation game well with the question outlined as the title of his text. He revealed the philosophical focus of his attack by writing:

> Before proposing my experiment, a few comments are in order. According to Aristotle, the property which properly distinguishes man from the rest of the universe is possession and use of the faculty of reason. If one may read between the lines of certain writings on the part of some philosophers who have been termed *Therapeutic Positivists*, Aristotle's distinction requires modification; man is unique by virtue of the ability to *mis*use the faculty of reason. All philosophical activity prior to approximately 1933 was according to the Therapeutic Positivist (hereafter referred to as T.P.) due to the misuse of reason. To be more explicit, philosophers reasoned incorrectly about the use of ordinary language; this resulted in the assertion of 'metaphysical' sentences.
>
> (Pinsky, 1951, p. 397, his emphasis)

This was, of course, a mockery of Wittgenstein.[36] Pinsky's mockery of the possibility of machine thinking concludes as follows. The machine is given Turing's 1950 paper to read, finds it stimulating, and has a thought, but then it is thinking about machine thinking. Since, according to the therapeutic positivist, this is the kind of puzzle or paradox that has led philosophy astray for centuries, the machine suffers a nervous breakdown.[37] Then, if the therapeutic positivist can convince the machine that it has been making 'metamechanical' statements, then 'the machine *does* think, since it has been able to misuse its thinking powers!' This, Pinsky suggested, 'is the experiment crucial' (1951, p. 398, his emphasis).

Two elements of this, which consists of the earliest reception of Turing, are noteworthy: first, Pinsky associates Turing's 1950 paper with Wittgenstein's philosophy and sees in Turing's work a similar threat to the one he seems to see in Wittgenstein's work; second, he sees 'Turing's game' as a proposal for a crucial experiment for the possibility of machine thinking.

Now, Mays referred to Turing's proposal as 'neo-behaviorism,' and he did not explicitly refer to Wittgenstein in his criticism of Turing. But it turns out that he was one of the participants in Wittgenstein's lectures at Cambridge University in the early 1940s, and it may be helpful to see how Mays remembered them:

> In his lectures, Wittgenstein made valiant efforts to quote examples to show that psychological data could be externalized. He talked a good deal about the criteria for deciding whether a person was in pain or not. Suppose, he said, so and so was on the operating table and surgeons were sticking knives into him; if he showed no signs of reacting, could he therefore be said to be in pain, or was he shamming? In these

examples, Wittgenstein sometimes tended to regard other people as if they were inanimate objects or automata, as when he said, 'Suppose I cut off Mr. X's arm thus,' at the same time, striking his own left arm with the edge of his right hand.

(Mays, 1967, pp. 83–84)

Mays criticizes both Turing and Wittgenstein under the same theme, namely, their alleged view of humans as (inanimate) machines. It is consistent with his earlier reference to Turing's 'neo-behaviorism' in (1952).

A couple of years later, from Princeton University in the United States, Claude Shannon (1916–2001) and his student John McCarthy (1927–2011) approached Turing to invite him to contribute a paper to their volume on cybernetics, which would be published in 1956, *Automata Studies* (1956).[38] Turing declined and wrote that his recent work has been on morphogenesis, though he expected 'to get back to cybernetics very shortly.'[39] In the preface of their book, McCarthy and Shannon thus received Turing's test proposal:

The problem of giving a precise definition to the concept of 'thinking' and of deciding whether or not a given machine is capable of thinking has aroused a great deal of heated discussion. One interesting definition has been proposed by A. M. Turing: a machine is termed capable of thinking if it can, under certain prescribed conditions, imitate a human being by answering questions sufficiently well to deceive a human questioner for a reasonable period of time. A definition of this type has the advantages of being operational, or, in the psychologists' term, behavioristic.

(McCarthy and Shannon, 1956, p. v)

McCarthy and Shannon received Turing's test as a 'definition' of intelligence, emphasizing the operational or behaviorist aspect of it as an advantage. However, they resumed and raised an objection to it:

A disadvantage of the Turing definition of thinking is that it is possible, in principle, to design a machine with a complete set of arbitrarily chosen responses to all possible input stimuli … Such a machine, in a sense, for any given input situation (including past history) merely looks up in a 'dictionary' the appropriate response. With a suitable dictionary such a machine would surely satisfy Turing's definition but does not reflect our usual intuitive concept of thinking. This suggests that a more fundamental definition must involve something relating to the manner in which the machine arrives at its responses—something which corresponds to differentiating between a person who solves a problem by thinking it out and one who has previously memorized the answer.

(McCarthy and Shannon, 1956, p. vi)

This objection states that a machine could be endowed with a large database containing answers to nearly all possible questions. The problem of passing the Turing test would then be reduced to the problem of how to collect, store, and retrieve fairly adequate answers to nearly all possible questions. From a conceptual point of view, machines of this kind can be understood as sophisticated mechanical parrots.[40] In any case, it is worth noting that McCarthy and Shannon considered the method by which the machine produced its behavior to be important. It had to resemble 'a person who solves a problem by thinking it out' rather than 'one who has previously memorized the answer.'

The shockwaves of Wittgenstein's philosophy and its possible influence on Turing do not appear in the North American reception of Turing. What does appear is the association of him with behaviorism, as McCarthy and Shannon put it, 'in the psychologists' term,' which is widely associated with the psychologist B. F. Skinner (1904–1990). Two of his influential books are *Behavior of Organisms* (1938), and *Verbal Behavior* (1957). In the first, Skinner sought to describe animal behavior using well-defined concepts such as stimulus, response, and reinforcement. He used highly controlled experiments with pigeons and rats, rewarding them with food in specific situations. In the second, Skinner sought to extend his approach to the study of verbal concepts as they appear in human language. As will be seen, most of the reception of Turing's 1950 paper after the 1950s was concerned with either attacks or defenses of him seen as a 'behaviorist' in various senses, but apparently no longer in connection with Wittgenstein. It is worth noting that in the cognitive sciences and philosophy of mind, the reputation of the behaviorist approach to pshychology was damaged after a critical review of Skinner's 1957 book written by Noam Chomsky in (1959). Chomsky captured a key point in Skinner's approach to analyzing behavior based on input-output functions, i.e., functional analysis, and generous quotes from him will pay off:

It is important to see clearly just what it is in Skinner's program and claims that makes them appear so bold and remarkable, It is not primarily the fact that he has set functional analysis as his problem, or that he limits himself to study of observables, i.e., input-output relations. What is so surprising is the particular limitations he has imposed on the way in which the observables of behavior are to be studied, and, above all,

the particularly simple nature of the function which, he claims, describes
the causation of behavior.

(Chomsky, 1959, p. 27)

Chomsky clearly positions his criticism as not against functional analysis per se, but against what he sees as a simplistic design of experimental conditions and the explanatory functions. He goes on to point
out what he finds lacking in Skinner's modeling, namely, the inclusion
of the internal structure of the organism whose behavior is to be predicted:

> One would naturally expect that prediction of the behavior of a complex
> organism (or machine) would require, in addition to information about
> external stimulation, knowledge of the internal structure of the organ-
> ism, the ways in which it processes input information and organizes
> its own behavior. These characteristics of the organism are in general
> a complicated product of inborn structure, the genetically determined
> course of maturation, and past experience. Insofar as independent neu-
> rophysiological evidence is not available, it is obvious that inferences
> concerning the structure of the organism are based on observation of
> behavior and outside events. Nevertheless, one's estimate of the rela-
> tive importance of external factors and internal structure in the deter-
> mination of behavior will have an important effect on the direction
> of research on linguistic (or any other) behavior, and on the kinds of
> analogies from animal behavior studies that will be considered relevant
> or suggestive.

(Chomsky, 1959, p. 27)

For Chomsky, the problem lied in Skinner's attempt to generalize results
from studies of simpler organisms such as pigeons and rats to the analysis of human verbal behavior. Chomsky emphasized that the difference in
the complexity between such organisms should be taken into account in the
design of the experimental study, which should consider the relative importance in each case of the internal structure of the organism and the external
stimuli:

> The differences that arise between those who affirm and those who deny
> the importance of the specific 'contribution of the organism' to learning
> and performance concern the particular character and complexity of this
> function [of how behavior varies in response to stimuli], and the kinds
> of observations and research necessary for arriving at a precise speci-
> fication of it. If the contribution of the organism is complex, the only
> hope of predicting behavior even in a gross way will be through a very

indirect program of research that begins by studying the detailed character of the behavior itself and the particular capacities of the organism involved.

(Chomsky, 1959, p. 27)

In light of Chomsky's review of Skinner, what can be learned about Turing's test proposal and its alleged connection to behaviorism?

On the one hand, Turing proposed a test that established certain controlled conditions. The interrogator, player C, should not see, touch, hear, or have any physical contact whatsoever with the participants, players A and B, the deceiver, and the supporter of the interrogator, respectively. The internal structure of the players was not open for analysis either. This experimental design, to the extent that Turing's test is taken as an experiment whose positive results alone can decide the issue, can certainly be seen as behaviorist. However, on the other hand, Turing devoted two sections of his paper (1950) to explaining the internal structure of digital computers, the specific kind of machine that he was considering in his argument. He outlined his belief that such a machine, given with sufficient memory and speed, and a suitable program should be able to imitate human verbal behavior well enough to deceive the interrogator. Turing thus acknowledged certain conditions on the internal structure of the organism being studied, in this case, the machine, in addition to the conditions of the experiment itself, arguably in the way that Chomsky wished that Skinner had one. What is the relative importance that Turing gave to each kind of condition? Did he prioritize experimental conditions for machine intelligence regardless of the internal structure of the main participant in the experiment, the machine? Chapters 3, 5, and 6 suggest that Turing significantly emphasized the latter over the former, so that Chomsky's criticism of Skinner will hardly apply to Turing.

Finally, it is worth noting in the late 1950s the reception of Michael Polanyi, who was a relatively close contemporary of Turing in Manchester and must have been one of the first readers of Turing's 1950 paper. But there is no record of how he received it early on. We only have his comments from *Personal knowledge* (1958), where he wrote:

I dissent therefore from the speculations of A. M. Turing (1950, p. 433) who equates the problem: 'Can machines think?' with the experimental question, whether a computing machine could be constructed to deceive us as to its own nature as successfully as a human being could deceive us in the same respect.

(Polanyi, 1958, p. 277)

Chapter 4 discusses some of the lively exchanges between Turing and Polanyi in Manchester in 1949, whose echoes can be seen in Turing's paper and his proposal of his test. For the purposes of this chapter, it is worth noting that Polanyi understood Turing's test proposal as an experimental question, rather than as a question of language and the meaning of words. This will be contrasted with Chomsky's explicit commentary on Turing's *Mind* paper in his own *Mind* paper, 'Language and Nature' (1995), as we will see shortly.

2.2   1960s–1990s: minds, machines, and consciousness

Still in the late 1950s, Ernest Nagel and James R. Newman published a book, *Gödel's Proof* (1958), that would influence the discussion of minds and machines for the next decade. Nagel and Newman aimed to interpret and disseminate Gödel's highly technical incompleteness theorems (Gödel, 1931), but they went on to a more open field. They claimed that 'the theorem [Gödel's proof] does indicate that the structure and power of the human mind are far more complex and subtle than any non-living machine yet envisaged' (p. 10). In the words of Turing's biographer Andrew Hodges (2008), '[b]y contrast with their detailed explanation of Gödel's technical arguments, they found no difficulty in writing off, in a few sentences, the possibility of artificial intelligence (AI),' even though 'Turing himself, in 1950, argued that Gödel's proof was irrelevant to the question of achieving AI.'[41]

An important discussion took place in the philosophy of mind at that time (Hook, 1960), and Hilary Putnam's contribution (1960) was structured around Gödel's theorems and Turing machines. He used the concept of a 'Turing machine' to argue that '[t]he various issues and puzzles that make up the traditional mind-body problem are wholly linguistic and logical in character' (p. 362). Putnam addressed two aspects of the mind-body problem, the question of privacy, or the meaning and significance of introspective reports of experience, and the question of the mind-body identity, or how mental states correlate with material and behavioral observables.

Regarding privacy, he argued that the question 'How do I know I have a pain?' is a deviant ('logically odd'), but the question 'How do I know Smith has a pain?' is not. Then, he stated that the same problem is mirrored in the case of a Turing machine T and its neighbor Turing machine T'. Regarding the identity problem, Putnam compared the relationship between internal states and external observables in humans and in machines, for example, a human being in pain and having C-fibers stimulated, and a Turing machine being in a given state 'A' when, and only when, flip-flop 36 is on. Essentially, perhaps oversimplifying Putnam's argument a bit, since the two problems are similar, but the temptation of dualism does not arise in the latter, he argued that dualism should not be a plausible solution to the former either.

Putnam acknowledged that he might be accused of advocating a 'mechanistic' world-view in pressing such an analogy. In reply, he stated: 'If this means that I am supposed to hold that machines think, on the one hand, or that human beings are machines, on the other, the charge is false.' And 'If there is some version of mechanism sophisticated enough to avoid these errors,' he added, 'very likely the considerations in this paper support it.' On Nagel and Newman's interpretation of 'Godel's theorem,' Putnam wrote:

> It has sometimes been contended (e.g. by Nagel and Newman in their book *Gödel's Proof*) that 'the theorem [i.e. Gödel's theorem] does indicate that the structure and power of the human mind are far more complex and subtle than any non-living machine yet envisaged' (p. 10), and hence that a Turing machine cannot serve as a model for the human mind, but this is simply a mistake.
>
> (Putnam, 1960, p. 366)

In fact, Putnam argued for using the Turing machine as a model for the human mind, specifically as a model to support analogical reasoning from the machine (known) to the mind (unknown).

But some philosophers, such as John Lucas (1961), would insist that Gödel's theorems had established that the power of the human mind exceeds that of machines, and therefore, there is no point in comparing the two.[42] According to Lucas (1961), in their private correspondence, Putnam would have 'suggested that human beings are machines, but inconsistent machines' (p. 120). Putnam may not have quoted Turing, but this is reminiscent of Turing's original line of argument against what he called ten years earlier 'the mathematical objection' (1950, p. 444), although Turing never stated his views so bluntly.[43]

As one of nine objections to the possibility of machine intelligence formulated and addressed by Turing, the mathematical objection argued why Gödel's incompleteness theorems and related work by others, including Turing's own paper (1936), should not be taken as implying the impossibility of machine intelligence. Before commenting on it, let us see Lucas' response to Putnam's view that humans are like inconsistent machines:

> Nor could we make its [the machine's] inconsistency a reproach to it — are not men inconsistent too? Certainly women are, and politicians; and even male non-politicians contradict themselves sometimes, and a single inconsistency is enough to make a system inconsistent. The fact that we are all sometimes inconsistent cannot be gainsaid, but from this it does

not follow that we are tantamount to inconsistent systems. Our inconsistencies are mistakes rather than set policies.

(Lucas, 1961, pp. 120–121)

This suggests that Lucas' rejection of Putnam's argument, and probably of Turing's own, was loaded with some values and expectations about the intelligence profiles of machines, men, women, politicians, and male non-politicians.[44]

In fact, Turing (1950) considered the possibility of making mistakes crucial for a machine to be able to develop intelligence.[45] He sought after a brain-inspired approach to achieving machine intelligence as a result of machine learning. He emphasized the connection between learning and fallibility, writing: 'Processes that are learnt do not produce a hundred per cent. certainty of result; if they did they could not be unlearnt' p. 459).[46]

Now, what about mechanism itself and the views that Putnam claimed not to hold, namely, that machines can think and that humans are machines? This was actually Lucas' target — he tried to prove, based on 'Gödel's Theorem,' that 'Mechanism is false, that is, that minds cannot be explained as machines.' He aimed his argument to be set out, 'with all objections fully stated and properly met' (p. 112), and the first such objection cited was Turing's 1950 paper (n. 2). Lucas takes for granted that Turing would have proposed that 'Mechanism' is true or that minds can be explained as machines.

Could it not be that Turing was looking for something that could satisfy Putnam's words, 'some version of mechanism sophisticated enough' to avoid the errors of entertaining that 'machines think' and that 'human beings are machines'? Although a study of Turing's concept of mechanism deserves a book-length treatment in its own right, Chapter 3 suggests that this question can be answered in the affirmative, i.e., the Turing test argument can best be read as saying that machines can *think$_2$*, and humans can *partially* be described as a kind of machine.

It is worth noting that, since Turing's machines (1936) and his paper 'Computing Machinery and Intelligence' (1950), the discussion of mind and mechanism has been centered on arguments for and against seeing the human mind as a Turing machine.

From the mid-1960s to the 1980s, the discussion continued with many interventions that are well reviewed by Shieber (2004), namely, the articles of Keith Gunderson (1964b), Richard Purtill (1971), Geoffrey Sampson (1973), P. H. Millar (1973), James Moor (1976), and Douglas F. Stalker (1978). These consist of criticisms and defenses of Turing's test as an experiment to determine the existence of an intelligent machine. Turing's test proposal was widely treated in philosophy as an essentialist definition of intelligence and in particular as the proposal of a sufficient condition for determining machine intelligence. This was despite Turing's own writings to the contrary.[47]

Fast forward to the early 1980s, and the charges of behaviorism against Turing reappeared in philosophy, now dressed up in the thought experiments of John Searle (1980) and Ned Block (1981). John Searle stated that the 'Turing test is typical of the tradition in being unashamedly behavioristic and operationalistic' (p. 423). In his critique of Turing's test and what he called 'strong artificial intelligence,' Searle proposed his Chinese Room argument to conclude that symbol manipulation can never amount to language understanding, even if it appears to be so to an outside observer. Searle's argument was posed against 'the claim that the appropriately programmed computer literally has cognitive states and that the programs thereby explain human cognition' or against 'any Turing machine simulation of human mental phenomena' (p. 417). Similarly, Ned Block wrote: 'Turing, for example, was willing to settle for a "sufficient condition" formulation of his behaviorist definition' (pp. 15–16). Block's answer came through a fictional memorizing machine, which he seems to associate with his aunt, Bertha. Presenting variants of McCarthy and Shannon's objection using thought experiments, Searle and Block also argued against the sufficiency of the Turing Test for determining intelligence. Viewing Turing's test as the proposal of a crucial experiment to determine the existence of mental states or consciousness in machines, they argued that passing the test would not guarantee anything of the kind.

A couple of years after Searle and Block's criticisms appeared, biographer Andrew Hodges weighed in. He wrote that Turing 'introduced the idea of an operational definition of "thinking" or "intelligence" or "consciousness" by means of a sexual guessing game' (1983, p. 415).[48] The behaviorist reading of Turing's proposal would persist. In 1990, Robert French would still read the Turing test 'as a simple operational definition of intelligence:'

> There are those who believe that passing the Turing Test constitutes a sufficient condition for intelligence and those who do not. The philosophical importance of this first claim is that it provided a clean and novel test for intelligence that neatly sidestepped the vast philosophical quagmire of the mind-body problem. The philosophical claim translates elegantly into an operational definition of intelligence: whatever *acts* sufficiently intelligent *is* intelligent.
>
> (French, 1990, p. 53)

Here we complete a full chronological arc from Mays (1952) to French (1990), much of which consists of reading Turing (1950) as a crassly behaviorist or operationalist philosophical paper. Turing would have considered the imitation of intelligence to be exhaustive of the concept of intelligence.

Recall Mays' interpretation that Turing would have considered intelligence a simple, directly measurable quality. Also, according to McCarthy and Shannon, Turing would have considered the method used by the machine to produce its behavior unimportant. The remaining chapters of this book will suggest that these readings are not very attentive to Turing's writings. Chomsky (1959) and Putnam (1960) are exceptions in that, while not specifically addressing Turing's 1950 paper in particular, they seem to offer more nuanced perspectives on the mind-body problem and behaviorism. Once juxtaposed with Turing's test proposal, their perspectives are not as reductive of it as the received view.

In Mays' postscript (2001) to his original critique of Turing in the early 1950s, he acknowledged Searle and noted that he was pleased with the change of fashion regarding the concept of consciousness:

> John Searle has pointed out that 'As recently as a few years ago if one raised the subject of consciousness in cognitive science discussions, it was generally regarded as a form of bad taste, and graduate students who are always attuned to the social mores of their disciplines, would raise their eyes to the ceiling and assume expressions of mild disgust.' There is now a greater tolerance to using the word consciousness. We even have a *Journal of Consciousness Studies*. But it would have been difficult for Ryle to admit that explanations of our higher mental processes in terms of consciousness might give us a better understanding of them than would explanations in behavioural terms. To do so would undermine his whole attack on the ghost in the machine. Nevertheless, Ryle's thought still lives on in the work of Daniel Dennet [*sic*] and others.
> (Mays, 2001, p. 4)

In fact, Daniel Dennett (1984) wrote what appears to be the most influential defense of the Turing test as a philosophical argument in the twentieth century.

Dennett noted that the test seemed to come from a longer philosophical tradition ('Perhaps he was inspired by Descartes,' p. 297). In any case, he claimed that it is capable of encompassing several specific intellectual tasks. So much so, he argued, that it can in fact be seen as a rather convenient sufficient condition (a 'quick probe,' p. 298) for determining the presence of human-level machine intelligence. Dennett complained that a 'failure to think imaginatively about the test actually proposed by Turing has led many to underestimate its severity and to confuse it with much less interesting proposals.' He went on to write: 'the Turing test, conceived as he conceived it, is (as he thought) plenty strong enough as a test of thinking.' And added: 'I defy anyone to improve upon it' (p. 297). Dennett made himself a champion of the Turing test, especially as a practical experiment for determining the existence of an intelligent machine. This would become even more visible in the developments of the early 1990s.

In the same year as Dennett's defense, another mockery appeared. Edsger Dijkstra (1930–2002), in a talk he entitled 'The threats to computer science' (1984), introduced a new fashion in computer science, the dismissal of Turing's test proposal as irrelevant to the field. Dijkstra was no minor figure (Hoare, 2003).He was a pioneer and leader in conceiving computer science as a discipline, and it was as part of that discipline that AI developed. In his 1984 talk, Dijkstra said:

> The Fathers of the field had been pretty confusing: John von Neumann speculated about computers and the human brain in analogies sufficiently wild to be worthy of a medieval thinker and Alan M. Turing thought about criteria to settle the question of whether Machines Can Think, a question of which we now know that it is about as relevant as the question of whether Submarines Can Swim.
>
> (Dijkstra, 1984)

Dijkstra acknowledged Turing and John von Neumann (1903–1957) as the founding 'fathers' of computer science, only to reject their original views and suggest that it was time for the field to turn its back on them.

2.3   The 1990s: aftermath of a 'social experiment'

The Loebner Prize competition was an event that was run annually for many years since 1991, offering a money prize to motivate programmers from academia, industry, and society in general to develop and submit their contestant machines to a real Turing test. The story runs as if this was the initiative of the North American inventor, industrialist and rich man, Hugh Loebner (1942–2016). However, a more accurate account of its origins was given by Loebner's friend, the psychologist Robert Epstein, a former student of B. F. Skinner (Section 2.1), who identified himself as the intellectual mentor behind the initiative:

> In 1985 an old friend, Hugh Loebner, told me excitedly that the Turing Test should be made into an annual contest. We were ambling down a Manhattan street on our way to dinner, as I recall. Hugh was always full of ideas and always animated, but this idea seemed so important that I began to press him for details, and, ultimately, for money. Four years later, while serving as the director of the Cambridge Center for Behavioral Studies, an advanced studies institute in Massachusetts, I established the Loebner Prize Competition, the first serious effort to locate a machine that can pass the Turing Test.
>
> (Epstein, 1992, p. 81)

One of Epstein's first considerations was to form a committee:

> The committee met every month or two for two or three hours at a time, and subcommittees studied certain issues in between committee meetings. I think it's safe to say that none of us knew what we were getting into. The intricacies of setting up a real Turing Test that would ultimately yield a legitimate winner were enormous. Small points were occasionally debated for months without clear resolution. Several still plague us.
>
> (Epstein, 1992, p. 81)

The academic distinction of the committee can hardly be questioned. Among its members, mostly scholars based in Cambridge, Massachusetts, Epstein recruited the historian of science I. Bernard Cohen (1914–2003), the computer scientist Joseph Weizenbaum (1923–2008), the philosopher Willard Van Orman Quine (1908–2000), and other distinguished scholars, including Daniel Dennett. The AI pioneer Allen Newell (1927–1992) served as an advisor.

According to Epstein, as a result of their two years of discussion, they decided to reject Turing's two-terminal design in favor of a one-terminal design that was 'more discriminating and less problematic' (p. 81). Thus, instead of three players, each instance of the imitation game would be played between player C, the human interrogator and judge, and either player A or B, that is, a machine pretending to be a human, and a human giving truthful answers to appear as human as possible. Epstein described at length the reasons why the committee ultimately decided to use a different version of Turing's test. The design for the 1991 edition was essentially that about ten judges would be faced with an equal number of terminals and would be told that at least two of the terminals would be controlled by computers and at least two by humans. Several other details make up their chosen design for the 1991 edition. They called it 'a restricted version of the classic Turing Test of machine intelligence,' where 'restricted' can be read as less powerful but feasible to implement. Each judge would spend about fifteen minutes at each terminal and score the terminal based on how human-like the session exchanges seemed to be. The positions would be rotated in a pseudo-random sequence. At the end, the players A, B, and C would have their success rates compared.

Although this implementation of Turing's test may seem overly simplistic from an armchair perspective, Epstein emphasized the challenges the committee faced in moving from interpreting Turing's 1950 paper to actually implementing his test proposal:

We spent months researching, exploring, and rejecting various rating and confidence measures commonly used in the social sciences. I programmed several of them and ran simulations of contest outcomes. The results were disappointing for reasons we could not have anticipated. Turing's brilliant paper had not gone far enough to yield practical procedures. In fact, we realized only slowly that his paper hadn't even specified an outcome that could be interpreted meaningfully. A binary decision by a single judge would hardly be adequate for awarding a $100,000 prize—and, in effect, for declaring the existence of a significant new breed of intelligent entities. Would some proportion of ten or even 100 binary decisions be enough? What, in fact, would it take to say that a computer's performance was indistinguishable from a person's?

(Epstein, 1992, p. 82)

They eventually adopted a 'simple scoring method' that led to what Epstein called a 'conceptual breakthrough:'

to have each judge rank the terminals according to how humanlike the exchanges were. The computer with the highest median rank wins that year's prize; thus, we are guaranteed a winner each year. We also ask the judges to draw a line between terminals he or she judged to be controlled by humans and those he or she judged to be controlled by computers; thus, we have a simple record of errors made by individual judges. This record does not affect the scoring, but it is well worth preserving. And, finally, if the median rank of the winning computer equals or exceeds the median rank of a human confederate [player B], *that computer will have passed (a modern variant of) the Turing Test.*

(Epstein, 1992, p. 82, his emphasis)

It is worth noting that this was the first systematic attempt to bring Turing's test proposal, until then an argument presented in a philosophy journal, into an experimental setting.

Thus the first edition of the Loebner Prize competition, or experiment, took place in 1991 at the Computer Museum in Boston. The winner was a Joseph Weintraub (1992, p. 90), then fresh graduate in psychology turned computer programmer. His program was called 'PC Therapist.' It was inspired by 'ELIZA,' a trick developed in the mid 1960's by Joseph Weizenbaum (1966), who was on Epstein's committee, to imitate a 'person-centered' therapist by turning one's own words and phrases back at them.

The Loebner competition continued to run annually for many years until recently, being organized by other scholars, and since 2014, in the

United Kingdom, when the first claim to pass the Turing test was made.[49] Still in its early years, the initiative drew analysis and criticism from computer scientists.

Stuart Shieber presented what appears to be the most rigorous and comprehensive analysis of practical Turing tests (1994a; 1994b), recently revisited and updated (2016). For practical Turing tests to be consistent with Turing's words, Shieber had noted early on (1994a), they cannot be restricted in domain (must be open to any conversation topic) or task (must be open to any question). Shieber recommended that practical Turing tests should not be run until the standard of AI comes close to the high standards required by the test. Shieber (1994a) made a comparison with the Kremer Prize for human-powered flying inspired by the designs of da Vinci. A cash prize was an appropriate incentive in that case, Shieber noted, because 'the task [was] just beyond the edge of current technology' (p. 74). He noted that 'limited tests are better addressed in the near term by engineering (building bigger springs) than science (discovering the airfoil)' (p. 77) and argued that in the case of the Turing test, there still is a substantial scientific gap to be filled. Years later, Shieber published his anthology on the philosophical reception of Turing's proposal (2004).

Judith Genova (1994), completely disconnected from the ongoing discussions of practical Turing tests, contributed what seems to be the best analysis of Turing's 1950 paper at the time. Even Hodges, who had spent over 500 pages of his Turing biography emphasizing Turing's homosexuality, had erased the question of gender from Turing's 1950 paper. But Genova, very attentive, instead of finding it a silly distraction, was able to connect it to the problem of evolution and natural kinds, understanding that Turing was rebelling against a fixed view of nature:

> Turing's passion structures the very character of his thinking; it is the critical factor that makes his ideas so fresh and exciting and allows him to challenge so many of Western culture's most cherished distinctions, e.g. that between the animate and inanimate, the real and the artificial, male and female. The plasticity of sex/gender and the social/cultural roots of thinking are innovative ideas (whether true or false) that owe their birth to his inability to keep his personal life out of his scientific one.
>
> (Genova, 1994, p. 324)

Perhaps following Hodges (1983), Genova sometimes psychologizes Turing in her writing, and she also often suggests that for Turing computing would be committed only to simulation and would not hold up a mirror to nature:

[T]he test of machine intelligence for Turing is not dialectical or discriminatory. To the extent that what we have traditionally meant by knowledge is an ability to discern or identify what independently is, I believe Turing to be proposing a new or different conception of knowledge and thinking. ... Unlike dialectic, which is intent on giving everything its proper place, computing is uninterested in the proper and thus holds no mirror to nature. Build it anyway you like as long as it works, i.e. as long as it emulates the behavior or the properties of the original. So-called natural differences are irrelevant to this process. The differences between apples and oranges, for example, or those between men and women, disappear in the realm of numbers. ... Good simulation collapses the distinction between the real and the simulated.

(Genova, 1994, pp. 319–320)

Against the view that Turing would have fallen prey to a fully relativistic account of knowledge and abandoned dialectic, Chapter 6 shows how Turing relied significantly on dialectic in structuring the argument of his 1950 paper.

Overall, Genova suggested that it must have been natural for Turing, as a homosexual, to believe that physical or biological constitution does not determine behavior. She noted: 'Indeed, the fact that gender is a matter of knowledge suggests that both thinking and being for Turing are discursive, cultural phenomena, not biological ones.' Rather, she concluded, 'to put the point more succinctly for Turing, biology is open to thought's manipulation' (p. 315). Although this is indeed a very perceptive reading of Turing's philosophy, Genova downplayed the seriousness of Turing's views on biology and (natural) evolution from the point of view of natural science, which are in continuity with his views on society and culture, including the plasticity of gender and thought. This is discussed in Chapter 7. In any case, Genova's analysis still appears to be by far the best available up to the early 2000s.

Chomsky, also a stranger to the ongoing discussions of practical Turing tests, would publish an influential paper in *Mind* (1995) that included an explicit comment on Turing's test proposal. Chomsky quoted from John Haugeland's commentary on AI (1979): 'how one might *empirically* defend the claim that a given (strange) object plays chess?' (p. 620, Haugeland's emphasis). Many of these debates 'over such alleged questions as whether machines can think,' Chomsky wrote (p. 9), 'trace back to the classic paper by Alan Turing.' They fail to take note, he claimed, that Turing himself declared to believe that the question 'can machines think?' was 'too meaningless to deserve discussion.' Chomsky concluded:

> It is not a question of fact, but a matter of decision as to whether to adopt a certain metaphorical usage, as when we say (in English) that airplanes fly but comets do not ... Similarly, submarines set sail but do not swim. There can be no sensible debate about such topics; or about machine intelligence, with the many familiar variants.
>
> (Chomsky, 1995, p. 9)

Chomsky caught the same metaphor about whether submarines can swim that Dijkstra had used ten years earlier. Chomsky emphasized an important aspect of the problem, namely that language adoption is a social phenomenon. But he seems not to have noticed what Genova was able to notice, namely that Turing's argument was more ambitious than a question about a certain metaphorical use of language. Chomsky denied Turing's question a place in the empirical sciences.

In March 1995, Marvin Minsky, who apparently never took the Turing test as a practical experiment (Section 1.2), hoped that Loebner would 'revoke his stupid prize, save himself some money, and spare us the horror of this obnoxious and unproductive annual publicity campaign.'[50]

Months later, in August 1995, Patrick Hayes and Kenneth Ford delivered a keynote address at the major AI conference of the time, the International Joint Conferences of AI (IJCAI). They presented a fierce critique of Turing's imitation tests, probably the most comprehensive collection of criticisms of the test as a practical experiment. Their argument can be seen as a development and extension of a point that was made earlier by Dijkstra in his 1984 lecture (Section 2.2). Not only had Hayes and Ford paraphrased a famous epigram of Dijkstra's from 1968 in the title of their paper — 'Turing test considered harmful' —, but also the Chomsky variant (Can airplanes fly?) of Dijkstra's ingenious metaphor. It is worth taking a moment to consider at the scope of their criticism.

They acknowledged that Turing's test 'has been with AI since its inception, and has always partly defined the field' (p. 972). Moreover, they recalled, '[s]ome AI pioneers seriously adopted it as a long-range goal, and some long-standing research programs are still guided by it.' This led them to set out to 'take Turing seriously' (p. 972). And they were assertive. In their view, Turing was *not* 'being merely metaphorical or speaking in some loose, inspirational way.' He was proposing the imitation game, they sensed, 'as a definite goal for a program of research.' It seemed to them that the test 'was supposed to be a concrete and relatively well-defined goal,' proposed rather 'to avoid the philosophical quagmire that Turing (correctly) predicted would result from debates about whether a computer could properly be described as "intelligent."' However, they went on to argue, the test has many ambiguities, flaws, and gaps in its design. For one thing, they pointed out, the test is based on the false assumption that intellectual talent must correlate with conversational skill or debating ability; for another, the 'gender' test — one of the possible interpretations of the test in which the machine takes the place of a man in trying to imitate a woman —, they went on to write, is not a test of making an 'artificial human' but of making a 'mechanical transvestite' (p. 973). In summary, they claimed that the test can't detect anything, is an elusive standard, has biases, is subject to gaming, and is even circular. 'Turing's dream,' they urged, ought not to be 'Turing's ghost' (p. 976, where they associated Turing and his test proposal with Mary Shelley's Frankenstein character). Claiming to have taken Turing's test seriously, Hayes and Ford sentenced:

> It had a historical role in getting AI started, but it is now a burden to the field, damaging its public reputation and its own intellectual coherence.' We must explicitly reject the Turing Test in order to find a more mature description of our goals; it is time to move it from the textbooks to the history books.
>
> (Hayes and Ford, 1995, p. 974)

Although Hayes and Ford's lecture should not be overrated, it can be seen as marking the moment when the Turing test was explicitly rejected, if not ridiculed, within the AI scientific community.[51]

Hayes and Ford's critique can be seen together with Shieber's analysis and Minsky's commentary as part of the aftermath of the Loebner Prize competition, which itself was seen in computer science and AI as ridiculing Turing's test proposal. However, unlike Shieber and Minsky, Hayes and Ford (1995) did not distinguish the lessons of the Loebner contest from Turing's original proposal in its historical context. In fact, they took such outcomes as evidence of the weakness of it:

> The Loebner competition illustrates very clearly how the imitation game inevitably slides from a concern with cognitive status to being a test of the ability of the human species to discriminate its members from mechanical imposters.
>
> (Hayes and Ford, 1995, p. 974)

We can now return to Daniel Dennett, who has not only been a member of the committee appointed by Epstein since 1991, but also chaired it for the first three editions. Would Dennett agree with Hayes and Ford's comment about the Loebner competition? Yes, he probably would. Since his eloquent defense of the power of the Turing test as a practical experiment for AI (a 'quick probe'), his experience with the Loebner contest seems to have changed his mind. A couple of years later, he wrote a postscript (1997).

Describing his experience with the competition, he reported that the quality of the contestants was low (p. 315). After a third edition he recommended adding an eliminatory phase before to the actual tests. The machine contestants would have to pass some specific natural language processing tasks in order to be eligible to participate in the variant of Turing's test. The goal was precisely to eliminate the 'mechanical imposters' that Hayes and Ford alluded to. Dennett's recommendations were not accepted. He resigned. Still in the postscript, he reflected:

> The Loebner Prize Competition was a fascinating *social experiment*, and some day I hope to write up the inside story … But it never succeeded in attracting serious contestants from the world's best AI labs. Why not? In part because, as the essay argues, passing the Turing Test is not a sensible research and development goal for serious AI. It requires too much Disney and not enough science … The Turing Test is too difficult for the real world.
>
> (Dennett, 1997, p. 315, my emphasis)

The passage is striking. As a philosopher who was willing to learn from experience and not to regard an honest change of mind as a sin, Dennett was able to appreciate the lessons of a social experiment. Rather than blaming Turing's 1950 philosophical paper, Dennett seems to have understood that, in fact, the Turing test may not really have been designed to serve as a practical experiment.

2.4   The 2000s: a new wave of commentary

Not all philosophers of mind and AI would accept to abandon the Turing test as a practical experiment. James Moor (1976) had earlier presented a defense of the Turing test as a form of inductive probabilistic inference: above a certain threshold of favorable accumulated evidence, there would be empirical support for concluding that machines can think or have thoughts, as ordinarily understood. While Moor has the merit of acknowledging the presence of a theoretical and an empirical component in Turing's scientific philosophy, it attributes to Turing an epistemological criterion that is reminiscent of the epistemology of Rudolf Carnap (1891–1970). Not only is Moor's interpretation of Turing ahistorical, it is also inconsistent with the historical placement of Turing in the context of his Cambridge milieu of the 1930s, as previously identified by Juliet Floyd (2017). The contributions of this book, particularly Chapters 5 and 6, will suggest that the impact of Wittgenstein's thought experiments, fictional dialogues, and anti-positivist philosophy was indeed felt by Turing.

Now in the early 2000s, Moor edited an anthology, *The Turing Test: The Elusive Standard of Artificial Intelligence* (2003). Three works are worth commenting in the context of this book: Jack Copeland (2000b), Gualtiero Piccinini (2000), and Susan Sterrett (2000).

Jack Copeland presented a defense of Turing's 1950 paper against the idea that Turing had proposed a definition for the concept of intelligence. But he also tried to dissociate Turing's test proposal from the criticisms that followed the Loebner competition, especially the claims that Turing was ambiguous and vague about the specifics of his alleged experiment:

It is often claimed that Turing was insufficiently specific in his description of his test. What are the specifications of a definitive test? How long? How many judges? What number of correct identifications is to be tolerated? However, these demands appear to miss the point. Whether a given machine is able to emulate the brain is not the sort of matter that can be settled conclusively by a test of brief duration. A machine emulates the brain if it plays the imitation game successfully come what may, with no field of human endeavour barred, and for any length of time commensurate with the human lifespan.

(Copeland, 2000b, p. 530)

Copeland then resumed suggesting what he thought was a way out:

Consider two time-unlimited imitation games, a man-woman game and a machine-human game, each employing the same diversity of judges that one might encounter, say, on the New York subway. If, in the long run, the machine is identified correctly no more often than is the man in the man-woman game, then the machine is emulating the brain. Any test short enough to be practicable is but a sampling of this ongoing situation. After some amount of sampling, we may become convinced that, in the long run, the machine will play as well as the man, but only because we believe that our samples of the machine's performance are representative, and we may always change our opinion on the basis of further rounds of the game.

(Copeland, 2000b, p. 530)

Copeland seems to believe that he can anticipate from the armchair the problems, surprises, and challenges that one faces in actually designing and running an experiment. He alluded to the use of sampling, which is not suggested in Turing's own text, but might be a valuable addition for those interested in implementing practical Turing-inspired tests. Although he thus far never pursued the proposal of any specific statistical sampling scheme, it turns out that Shieber did, as we will see.

Since Copeland defends an interpretation of the Turing test as a practical experiment for AI, the Loebner Prize competition might be the kind of initiative he would like to see implemented, revised, improved, etc., instead of turning a blind eye to it. Such a coherence is shown by Moor, who insisted on his interpretation of the Turing test as a practical experiment to accumulate evidence for inductive probabilistic inference (2001), but took the trouble to comment on the Loebner competitions (*Ibid.*) and even later collaborated with them (Warwick et al., 2013). The pressure that the Loebner Prize competition puts on philosophers such as Copeland and Moor will only increase in the 2010s, when a chatbot that is widely considered unintelligent by computer scientists and AI researchers as unintelligent is announced to have passed a practical Turing test (Vardi, 2014; Marcus et al., 2016), and the experiment is claimed to have been implemented by strictly following Turing's words (Warwick and Shah, 2015). Chapters 5 and 6 discuss these developments.

Piccinini (2000) aimed at discussing 'the rules' of Turing's test proposal, which is a term Turing never used in this sense. This provides an opportunity to illustrate the kind of methodological abuse that Castañeda criticized in the history of philosophy (Section 1.5.1). Instead of engaging in the exegesis of each source, and then perhaps later looking for points of connection between different sources, Piccinini employs the Athenian approach to read Turing's 1950 paper as a unit within the larger corpus of Turing sources. Piccinini referred to the so-called 'standard' and 'literal' readings of Turing's imitation game and test as alternative and mutually exclusive. According to the former, 'the goal of the interrogator was to discover which was the human being and which was the machine, while the goal of the machine was to be indistinguishable from a human being.' According to the latter, 'the goal of the machine was to simulate a man imitating a woman, while the interrogator — unaware of the real purpose of the test — was attempting to determine which of the two contestants was the woman and which was the man.' Piccinini thus justified his conclusion in favor of the former reading:

> The present work offers a study of Turing's rules for the test in the context of his advocated purpose and his other texts. The conclusion is that there are several independent and mutually reinforcing lines of evidence that support the standard reading, while fitting the literal reading in Turing's work faces severe interpretative difficulties. So, the controversy over Turing's rules should be settled in favor of the standard reading.
>
> (Piccinini, 2000, p. 573)

Piccinini searched Turing's 1950 text and other primary sources for evidence in favor of each reading. He found more evidence in favor of the 'standard' reading, the winning interpretation of the dichotomous problem imposed into Turing's texts. Regarding the parts of Turing's text that the standard reading could not explain, however, Piccinini struggled with Turing's text and concluded:

> According to those who have witnessed or studied his life, Turing was often a surprisingly fast thinker. He would get frustrated when others took a long time to get points that seemed obvious to him. Perhaps because of this, his writing was lucid but not always easily understood … In light of this, the most likely explanation for the ambiguity in Turing's rules is that he expected his readers to fill in the details in accordance with the game's purpose … Turing's own imitation game did *not* involve a machine simulating a man who's pretending to be a woman, but a machine simulating a human being.
>
> (Piccinini, 2000, p. 580, his emphasis)

Turing, the philosopher and author under study, ends up being psychologized, while the interpreter can conclude with peace of mind. The question of gender in Turing's 1950 paper, which is part of what Piccinini calls the 'literal' reading, ends up simply being erased from Turing's philosophy.

Susan Sterrett (2000) can be read as arguing for the acknowledgement of Turing's 1950 historical imitation game and against the idealized Turing test that has been construed in philosophy, e.g. what Piccinini called its 'standard' reading. Instead of reading Turing's text from the point of view the achievements of contemporary AI and philosophy of mind, Sterrett offered a close reading of Turing's own text. Her interpretation appears to be the most accurate reading of Turing's 1950 proposal. She noted that gender impersonation requires what is one of the most difficult intellectual skills, that of thinking quickly but reflectively enough to respond to a new situation in a way that deviates from one's most deeply held experiences. Sterrett wrote: 'the significance of the use of gender in the [imitation game] is in setting a task for the man that demands that he critically reflect on his responses' (p. 550).[52]

Sterrett presented what appears to be by far the most perceptive analysis of Turing's test proposal, including the challenging interpretive problem of gender impersonation in it, without straying from a rigorous reading of Turing's text and psychologizing Turing, as Genova sometimes did. Sterrett noted that the machine-imitates-woman version of the imitation game (pp. 433–434) 'realizes a possibility that philosophers have overlooked: a test that uses a human's linguistic performance in setting an empirical test of intelligence, but does not make behavioral similarity to that performance the criterion of intelligence' (2000, p. 541):

> [T]he difficult task the man is set by the criterion used in the [man-imitates-woman variant of the game] requires in addition that stereotypes get used for different purposes: rather than serving as common background drawn on in sincere efforts to communicate, they are to be used to mislead someone else to make inferences to false conclusions, which requires more reflection upon how others make inferences than is normally required in conversation.
>
> (Sterrett, 2000, p. 547)

Sterrett's point is underscored by a crucial observation that she found in Gilbert Ryle's *The Concept of Mind*. The intellectual skills required for stereotype impersonation are perhaps most evident in the performance of a clown:

> The cleverness of the clown may be exhibited in tripping and tumbling. He trips and tumbles on purpose and after much rehearsal and at the golden moment and where children can see him and so as not to hurt himself ... The clown's trippings and tumblings are the workings of his mind, for they are his jokes; but the visibly similar trippings and tumblings of a clumsy man are not the workings of that man's mind.
>
> (Ryle, 1949, p. 33)

A corollary of Sterrett's point is that Turing's proposal can hardly be related to behaviorism. While behaviorism is based on a well-defined behavioral task to be evaluated in an experiment, the intellectual task of imitating a 'woman' is actually in question is slippery — what exactly is a stereotypical female behavior in a given culture? Overall, and also recalling Genova's interpretation (§2.3), Turing can be interpreted as critically addressing stereotypes of intelligent *v.* mechanical, female *v.* male, human *v.* machine, natural *v.* artificial, which he took to be rather fluid and embedded in a given culture.

The 2000s also saw the publication of the Shieber anthology mentioned above, including a new commentary by Chomsky. In this occasion, Chomsky appears to have been a bit more positive about Turing's test proposal. He emphasized that '[t]he dual significance of the enterprise [proposed by Turing] — constructing better machines, gaining insight into human intelligence — should no longer be in doubt, if it ever was' (p. 317). Chomsky praised the former and did not condemn the latter, but criticized it, approximating it with Wittgenstein's philosophy:

Turing said nothing more about why he considered the question he posed at the outset —'Can machines think?' — 'to be too meaningless to deserve discussion,' or why he felt that it would be 'absurd' to settle it in terms of 'common usage.' Perhaps he agreed with Wittgenstein that 'We can only say of a human being and what is like one that it thinks'; that is the way the tools are used, and further clarification of their use will not advance the dual purposes of Turing's enterprise.

(Chomsky, 2004, p. 318)

Chomsky was implicitly referring to aphorisms §359 and §360 of Wittgenstein's *Philosophical Investigations*,[53] published posthumously and after Turing's 1950 paper, even though nearly the same thoughts had appeared and circulated in Cambridge earlier *c.* 1993 as part of the *Blue Book* (Wittgenstein, 1935). Chomsky clearly echoes Wittgenstein's point in the aphorisms, once again denying the question of whether machines can think a place in the empirical sciences (see Section 2.2). He resumed:

One can choose to use different tools, as Turing suggested might happen in fifty years, but no empirical or conceptual issues arise. It is as if we were to debate whether space shuttles fly or submarines swim. These are idle questions. Similarly, it is idle to ask whether legs take walks or brains plan vacations; or whether robots can murder, act honorably, or worry about the future. Our modes of thought and expression attribute such actions and states to persons, or what we might regard as similar enough to persons.

(Chomsky, 2004, p. 318)

Chomsky illuminates the question, emphasizing that it depends on whether recourse to machine thinking would be of any use to us. But he concludes by overlooking the point that Turing was actually engaged in a dispute about the scientific and technological possibility of making machines intelligent enough for us to possibly consider them 'similar enough to persons.'

Also echoing the reception of Turing's test proposal in the 1980s, Shieber (2007) presented a statistical sampling scheme to show that the criticisms of Turing's test on the grounds of sufficiency behaviorism are not plausible. He drew upon the notion of interactive proof from theoretical computer science, along with a set of physical constraints in the information capacity of the test's participants. Inferring the presence of mental states, he argued, would depend on either inductive or abductive inference. But these forms of logical inference to mental states could be replaced by 'statistical proof under weak realizability assumptions.'

Shieber's elegant mathematical approach can be seen from the point of view of the philosophy of common sense. That is, it can be approximated to Wittgenstein's arguments about anti-solipsistic criteria in the *Blue Book*, e.g., for experiencing that someone else is in pain:

> In fact the solipsist asks: 'How can we believe that the other has pain; what does it mean to believe this? How can the expression of such a supposition make sense?' ... The grammatical difficulty which we are in we shall only see clearly if we get familiar with the idea of feeling pain in another person's body. For otherwise, in puzzling about this problem, we shall be liable to confuse our metaphysical proposition 'I can't feel his pain' with the experiential proposition, 'We can't have (haven't as a rule) pains in another person's tooth.'
>
> (Wittgenstein, 1953, pp. 48–49)

Seen from this angle, the result of Shieber's 'very slight weakening of the criterion from one of logical proof to one of statistical proof' is not so weak. It moves from the metaphysical to the experiential level of analysis, and thus connects with the arguments of Wittgenstein — and Turing. To see the connection with Turing, note that the interrogator and judge in the imitation game, who is supposed to be a non-expert on machines, must make a decision after *experiencing* — not controlling in the Baconian sense — some (unspecified) reasonable amount of time in conversation with the A-B players. The perception of an imitation is an experience.

In fact, it is a bit odd to imagine giving the *non-specialized* interrogators and judges envisioned by Turing, ultimately the representatives of common sense, a sophisticated statistical tool on which to base their decisions. As Shieber (2016) himself later noted, the Turing test 'works exceptionally well as a *conceptual* sufficient condition for attributing intelligence to a machine, which was, after all, its original purpose' (p. 95, my emphasis).

In fact, the question of the non-specialized profile of the interrogators and judges would reappear soon. Luciano Floridi et al. (2009) contributed with the observation that for the Turing test to be useful, the interrogator must ask challenging questions, such as 'if we shake hands, whose hand am I holding?,' or 'I have a jewellery box in my hand, how many CDs can I store in it?,' or 'if London is south of Oxford, is Oxford north of London?' All of these questions, they report, 'immediately gave away both humans and machines, making it unnecessary for any further interaction or tests' (p. 147). Of course, Floridi et al. were aware of the so-called Eliza effect long reported by Weizenbaum (1966), and they adopted a relatively simple strategy to neutralize it. They argued:

First, and especially given the very short interaction, answers should be as informative as possible, which means that one should be able to maximise the amount of useful evidence obtainable from the received message. It is the same rule applied in the 20 questions game: each question must prompt an answer that can make a very significant difference to your state of information, and the bigger the difference the better.

(Floridi et al., 2009, p. 147)

Chapter 5 discusses how this relates to Turing's original writings and their context. In particular, their reference to the Victorian parlor game *Twenty Questions* is strongly linked to Turing's inspiration from the material culture of the time. They continued:

Second, questions must challenge the syntactic engine which is on the other side. So other questions such as 'what have you been up to today?' or 'what do you do for a living?' (again, two real examples) are rather useless too. The more a question can be answered only if the interlocutor truly understands its meaning, context or implications, the more that question has a chance of being a silver bullet.

(Floridi et al., 2009, p. 147)

Floridi et al.'s argument would later reappear in computer science and AI under the guise of the so-called Winograd Schema Challenge,[54] which asks that verbal machine behavior be evaluated by questions that actually require common sense knowledge from the machine.

The context of Floridi et al.'s analysis is that they were invited by Kevin Warwick and Human Shah to participate as interrogators and judges in the Loebner Prize competition. Understanding that they were participating in a *scientific* experiment, they aimed to be rigorous.

They were surprised, they report, by the naivety of other participants, notably the non-expert journalists. However, Turing was quite explicit in his writings that the interrogators and judges should indeed be non-experts in machine intelligence. Chapter 5 will revisit this issue and address it more assertively. Overall, the experience of Floridi et al. can be seen as favorable evidence for the argument outlined in Chapters 5 and 6 that the Turing test is a *thought* experiment.

2.5 The 2010s: some of the latest commentary

Since the 2010s, more analyses of the Turing tests have appeared. Those that seem more relevant from the point of view of this book will be discussed in due course later, especially, in the analytical chapters. Two of them, namely,

Darren Abramson (2011) and Diane Proudfoot (2013), will be reviewed now.

Darren Abramson (2011) revisited the connection between Turing and Descartes, which has often been made before him in the philosophy of mind,[55] and tried to push it further. He claimed: 'Turing at least conceived of his own test as fulfilling just the epistemic purpose that Descartes' fulfilled for him' (p. 550). Indeed, there is some symmetry between Descartes' proposal of his 'language test' as an empirical basis for providing — unfavorable — evidence for 'the presence of some property that is necessary for mind' in machines and in other animals, on the one hand, and Turing's proposal of his test to provide — favorable — evidence of the same kind.

At a first glance, Abramson's research may appear to be contextualized in history, as he attempts to link Turing to Descartes through material evidence — he refers to Turing's personal offprint of Jefferson's article, whose Descartes quote Turing marked in pencil, and certainly makes a contribution. However, his insufficient appreciation of the historical debate in which Turing was involved may have led him to overestimate the extent to which Turing traveled into the past to seek dialogue with Descartes, as opposed to contemporaries such as Jefferson himself and colleagues from the cybernetics movement, for example, Grey Walter (1910–1977).[56] Abramson was eager to refute the reception of the Turing test in the philosophy of mind as a form of behaviorism (Section 2.2), and this biases his analysis. He concluded by projecting a dialogue between Turing and Descartes *in the terms* of late twentieth-century philosophy of mind, which is alien to both Turing and Descartes. He found it ironic that 'Turing, an apparent materialist about mind, and Descartes, a dualist, agree on how we can determine that machines do or don't have minds' (p. 548).

In continuity with an earlier work (2008), Abramson argued that Turing's test expresses his 'commitment to a necessary condition for thought.' Since this condition is based on an inner, non-behavioral property, Turing's proposal cannot be behaviorist. Abramson called the inner property 'the epistemic-limitation condition,' which Turing would have revealed 'in his answer to "Lady Lovelace's objection"' (p. 546). The condition is a representation of the relationship between the machine's display of intelligent behavior, on the one hand, and the knowledge of the person who designed it, on the other. If the actual behavior of a machine goes beyond what its designer could anticipate or predict, then the machine can be said to be thinking. The 'presence of a mind,' which Abramson attributes to Turing as a claim that his test would be able to establish is Abramson's own reification of the so-called 'epistemic-limitation condition.'

Diane Proudfoot (2011) introduced the idea that deception can be and should be explored and controlled for in AI experiments. But beyond that, she urged AI researchers to acknowledge the value of the Turing test as a practical experiment to determine machine intelligence. Soon later, Proudfoot (2013) went on to propose a new interpretation of Turing's test proposal. She argued to have identified in two of Turing's works (1948; 1952) the view that intelligence is emotional, rather than mathematical, which she associated with the concept of response-dependence.[57]

For Proudfoot, interpreting Turing, a machine can be said to be intelligent if it appears intelligent to 'a normal subject' in certain 'specified conditions' of observation (Proudfoot, 2013, p. 404). Soon later Proudfoot (2017b) would write that '*the Turing test does not test machine behaviour*' (p. 303, her emphasis). 'Instead,' she continued, 'it tests the observer's reaction to the machine.'

Response dependence can be illustrated through other secondary-quality concepts. For example, a color can be perceived similarly by people who are not colorblind in adequate lighting conditions. This should not in principle preclude a physics of color, which reifies color as a (response-independent) primary quality concept. However, Proudfoot commits to a notion of 'global response-dependence' (Pettit, 1991, p. 588). She takes 'the concept of colour' as being 'very different from the concept of electromagnetic radiation, even though electromagnetic radiation is the physical basis of colour.' 'Likewise,' Proudfoot (2017b) concludes, 'if intelligence is a response-dependent concept, the concept of intelligence is very different from the concept of computation, *even if* brain processes (implementing computations) form the physical basis of "thinking" behaviour' (p. 305, her emphasis). Proudfoot is committed to anti-physicalism: she rejects the reification of the physical concepts of color and intelligence as primary-quality concepts, neglecting the fact that concepts originally formed in technical disciplines may eventually be absorbed into common sense, e.g., the physical concept of force. She imputes to Turing the view that intelligence is a socially constructed concept whose verifiability rests on the intersubjective judgment of human interrogators. The inescapable conclusion is that if an unintelligent chatbot can fool humans under the given socially accepted conditions, then the chatbot can be called intelligent. Chapter 5 shows that this view is at odds with Turing's writings.

Michael Wheeler (2020) presents an excellent critique of Proudfoot's interpretation of Turing and his test. Wheeler emphasizes the point that Proudfoot's interpretation of Turing in terms of response dependence is ambivalent and falls prey to relativism. He concludes that there is a variant of response dependence, the *reference-fixing* one, which seems more consistent with Turing's views, and may be a promising way to reclaim the value of the Turing test as a practical experiment for determining machine intelligence. But Wheeler's argument is more a critical contribution to

Proudfoot's initiative than a positive proposal of its own. It does not attempt to go into the depth of the Turing sources, nor does it discuss the existing experiences with practical Turing tests, such as the Loebner Prize competition.

Although not explicitly mentioning practical Turing tests, the computer scientist Stuart Russell (2019) contributed an opinion on the Turing test that turns out to be a skillful and very perceptive reading of Turing's sophisticated philosophical text:

> The imitation game serves a specific role in Turing's paper — namely as a thought experiment to deflect skeptics who supposed that machines could not think in the right way, for the right reasons, with the right kind of awareness. Turing hoped to redirect the argument towards the issue of whether a machine could behave in a certain way; and if it did — if it was able, say, to discourse sensibly on Shakespeare's sonnets and their meanings — then skepticism about AI could not really be sustained.
>
> (Russell, 2019, ch. 2)

Stuart Russel's comment is remarkable in that he considers *what kind of philosophical argument* Turing developed in his 1950 paper, and his interpretation is indeed largely consistent with that presented in this book. He implicitly agrees with Minsky, and further notes that the Turing test is a thought experiment with the goal of, say, shifting the original question, the natural order, etc. (§1). He then clearly departs from earlier interpretations that Turing would have proposed his test as a kind of crucial practical experiment to determine the existence of machine intelligence:

> Contrary to common interpretations, I doubt that the test was intended as a true definition of intelligence, in the sense that a machine is intelligent if and only if it passes the Turing test. Indeed, Turing wrote, 'May not machines carry out something which ought to be described as thinking but which is very different from what a man does?'
>
> (Russell, 2019, ch. 2)

Finally, he summarized what can be seen as the widely shared view of the majority of AI scientists about the Turing test:

> Another reason not to view the test as a definition for AI is that it's a terrible definition to work with. And for that reason, mainstream AI researchers have expended almost no effort to pass the Turing test.
>
> (Russell, 2019, ch. 2)

Stuart Russel seems to suggest that the Turing test is of no value to AI research as a *practical* definition, or to put it another way, as a practical working experiment, which is very much in line with the contributions of this book.

2.6 Conclusion

In this history of the reception of the Turing Test, there are a few main conceptual elements to emphasize. In Turing's closest intellectual milieu, in Cambridge, his test proposal was well received by Bertrand Russell, and there is some evidence that it would probably have been positively received by Wittgenstein as well. I will leave here a suggestion, which will not be further developed in this book, that Turing's philosophy seems to have collected elements from both (the later) Wittgenstein's and Russell's philosophies, despite the differences between the philosophical frameworks of the latter.

Regarding the interpretation of Turing's test proposal as a form of behaviorism, leaving aside the politicized colors of Mays' critique, Chomsky's critique of Skinner may be an informative reference and benchmark against which to assess Turing's views. Chomsky emphasized the relative importance that is given to experimental conditions, as opposed to the internal structure of the organisms in Skinner's experiments. Although this will not be a central issue in the following chapters, their analyses may shed light on how Turing can be positioned in relation to Chomsky's question. The criticisms of thinkers such as Searle and Block, on the other hand, seem to have paid little attention to Turing's own writings, and seem rather to have relied rather on oversimplified views of Turing's imitation principle and his machine conceptions in order to advance their own arguments, in short, the so-called straw man fallacy. Pinsky, despite the sarcasm and short shrift, seems to be a much more attentive reader of Turing, correctly identifying him with Wittgensteinian therapeutic positivism. Lucas' playing of the Gödelian card, in turn, can be reduced to Polanyi's earlier criticism of Turing, which will be seen in some detail in Chapter 4. Putnam's 1960 analysis and distinctions, despite the later shifts in his own philosophy, seem surprisingly robust in the face of the contributions of this book, especially Chapter 3. The reason that Putnam's careful distinctions partially match Turing's language may be that Turing, despite his irreverent imitation game, did not completely dismiss the mind-body problem as a philosophical problem with an empirical component.

The practical Turing tests implemented in the context of the Loebner Prize competition have arguably been of great value for any reflection on the Turing test. But philosophers, with the exception of Dennett, Shieber, Moor, and Floridi, have paid little attention to them. The Turing test is claimed to be valuable as a practical experiment for AI research, and yet philosophers who support Turing's test proposal have paid little attention to the criticisms of AI scientists and practitioners. Chapter 5 formulates this as a sort of dilemma that can be solved by interpreting the Turing test as a thought experiment in the modern scientific tradition.

3   Turing's Imitation Principle

Turing's imitation test was widely received as a reductionist view of intelligence, as if he were suggesting that intelligence is a directly measurable quality. A recent counter-interpretation is that he understood intelligence as an 'emotional' (socially constructed) concept intended to test the human interrogators and judges, not the machines. Both the reductionist and constructionist views downplay Turing's theoretical construction of machine intelligence and confuse his principle of imitation with identity. Both interpret Turing's concept of intelligence as identical to either an operational definition or a socially accepted convention. Either way, it is as if Turing rejected a distinction between the imitation of behavior considered intelligent by human experimenters, interrogators, and judges, and intelligence itself as built into an autonomous agent. This chapter argues that Turing conceived machine intelligence (think$_2$) in analogy to human intelligence (think$_1$), not identity. But he did aim to construct think$_2$ to 'simulate the behaviour of the human mind very closely,' and he was curious to see if there was any intellectual behavior that would ultimately remain beyond the reach of machines. While he did not aim to reconstruct the human mind, he did aim to explore by analogy how mechanical the human mind could be.

3.1   Argument sketch

Turing suggested that, if a machine can successfully imitate human intelligence in a general intelligence task such as conversation, then the meaning of the words 'machine' (previously associated with repetitive tasks) and 'thinking' (previously associated with something only humans can do) should be expanded. Most interpreters, whether critics or supporters, have understood this in a way that amounts to collapsing the imitation of intelligence with intelligence itself, collapsing analogy with identity. This chapter claims that Turing's argument is more nuanced than previous interpretations suggest.

There are two main interpretations of Turing's concept of intelligence circulating in the secondary literature, which in short can be regarded as the reductionist and the constructivist interpretations of Turing's concept of intelligence.[58] Both see the Turing test as a practical experiment to determine the existence of an intelligent machine. The reductionist view is that Turing reduced intelligence to a certain form of verbal behavior

DOI: 10.4324/9781003300267-3

that fits an operational definition. This is often followed by the assumption that machine behavior is equivalent to symbol manipulation by a Turing machine, which is supposedly how Turing thought the human mind works. Alternatively, the constructionist view holds that Turing designed his imitation game to test the human perception of machine intelligence, not machine intelligence itself. According to this reading of Turing, no mathematical theory of intelligence is possible, because intelligence is actually an elusive concept whose meaning can only be established by convention, by social agreement on certain psychometrics.

This chapter argues that these two views of Turing's concept of intelligence are no more than cartoons.[59] First, I refer to the main primary sources to trace the role of machine learning and imitation in Turing's construction of machine intelligence. I suggest that Turing conceived of machine learning as the theoretical foundation for machine intelligence (think$_2$), and that he conceived of it as a mathematical and computational science (Section 3.2). I also explore Turing's search for an empirical basis for machine intelligence, an approach that could neutralize confirmation bias against the machine (Section 3.3). I further suggest that he drew inspiration from human intelligence (think$_1$), but did not aim to literally reconstruct it in the machine. Rather, he considered the *imitation* of think$_1$ as the empirical basis for think$_2$ (Section 3.4), in part because he was compelled to do so by specific challenges posed by contemporaries.

Showing in the primary sources that Turing's scientific philosophy had both a *theoretical* and an *empirical* basis for machine intelligence should be enough to support my argument that Turing did not collapse the imitation of human intelligence with human intelligence itself, that he did not conflate analogy with identity, as implied by the reductionist and the constructivist readings of his concept of intelligence. But I will go further and argue that:

(i) Turing considered the exact reconstruction of think$_1$ to be an intractable problem, and that
(ii) he considered the construction of think$_2$ to imitate think$_1$ itself to be quite a challenge.

Finally, despite these deflationary points, I further argue that Turing was anything but an instrumentalist — he pursued his research *not* for practical purposes, but rather to improve our understanding of nature *and* society.[60] While he did *not* aim to reconstruct the real human mind literally, he did aim to explore the extension and limits of mechanism as constitutive of the human mind, i.e., how mechanical the mind could be (Section 3.5). He did expect the use of words to shift in everyday life and common sense in the presence of intelligent machines, he did expect his science to inform our interpretation of the human mind. I conclude by outlining key points of Turing's scientific philosophy (Section 3.6).

3.2   Learning is the foundation for think$_2$, 1945–1952

Turing's image of intelligence consists, in his words, of 'learning like a brain' and 'learning from experience,' and these two phrases evoke two aspects of his views on learning and intelligence.

On the one hand, I suggest that by 'learning like a brain' Turing meant not just a behavioral disposition, for the learning should be grounded in the machine in the form of *changes in its instruction tables,* just as it does in the human brain as changes in its neural structure. Note that this establishes the nature of think$_2$ as intrinsically different from think$_1$, and Turing's explicit references to the grounding of think$_2$ in digital computer architecture, as we shall see shortly, suggest his awareness of this difference. At the same time, despite of this conceptual grounding, in contrast to dualistic philosophies of mind, Turing did *not* regard the brain as the hardware of the mind. Rather, he focused on the topological and functional properties of the brain, seen as a form of software.[61] He sought to understand the brain's intelligence power in terms of a grounded disposition for learning and development. In (1948), Turing referred to the 'intellectual power' of humankind and other animal species (p. 410). On the other hand, I suggest that by 'learning from experience' he considered the experiential and social dimension of learning. In (1948), he wrote: 'The isolated man does not develop any intellectual power.' He added: 'It is necessary for him to be immersed in an environment of other men, whose techniques he absorbs during the first 20 years of his life' (p. 431).[62] Turing's colleague Donald Michie (2002) once said: 'Turing's belief about intelligence was that the PROPENSITY is INNATE, but the ACTUALITY has to be BUILT' (his emphasis).

This section will show that Turing's identification of machine learning as the theoretical basis for machine intelligence (think$_2$) persisted over a chronological arc from at least late 1945 to early 1952.

3.2.1   *December 1945: a machine can display intelligence if it is allowed to make mistakes*

At least as early as his December 1945 report to the National Physical Laboratory (NPL), Turing referred in print to machine 'intelligence,' which in his view was possible should the machine be allowed to make mistakes:

'Can the machine play chess?' It could be fairly easily be made to play a rather bad game. It would be bad because chess requires intelligence … There are indications … that it is possible to make the machine to display intelligence at the risk of its making occasional serious mistakes. By following up this aspect the machine could probably be made to play very good chess.

(Turing, 1945, p. 389)

This connection between intelligence and error was not obvious at the time and indeed turned out to be crucial in Turing's evolving concept of machine intelligence. Later in 1950, he would add crucial insight into it:

> Another important result of preparing our machine for its part in the imitation game by a process of teaching and learning is that 'human fallibility' is likely to be [mimicked] in a rather natural way, *i.e.*, without special 'coaching' ... Processes that are learnt do not produce a hundred per cent. certainty of result; if they did they could not be unlearnt.
>
> (Turing, 1950, p. 459, no emphasis added)

This passage establishes how Turing related intelligence and error through learning. Turing's reference to 'mistakes' in December 1945 can be seen as a first milestone in Turing's view of machine intelligence as based on learning.[63]

3.2.2   November 1946: learning like a brain means changing logical structure in response to stimuli

In about November 19, 1946, Turing replied to a letter from the psychiatrist and cyberneticist Ross Ashby (1946) in which he discussed the Automatic Computing Engine (ACE), the digital computer whose design he was leading at the NPL. Referring to his project to make the ACE imitate the brain, he stated that he was 'more interested in the possibility of producing models of the action of the brain than in the practical applications of computing.' He went on to distinguish two kinds of 'uses' of the ACE for the imitation of the brain, one imitating its 'lower centres' and the other imitating its 'higher centres.' This is how Turing described the first kind, which would be similar to 'the action of the lower centres' of the brain:

> The ACE [as a universal Turing machine] will be used, as you suggest, in the first instance in an entirely disciplined manner, similar to the action of the lower centres [of the brain], although the reflexes will be extremely complicated. The disciplined action carries with it the disagreeable feature, which you mentioned, that it will be entirely uncritical when anything goes wrong. It will also be necessarily devoid of anything that could be called originality.
>
> (Turing, 1946)

Turing explained that there is nothing in the ACE's construction that obliges us always to use it in that way. He then moved to the second kind of action of the brain:

There is, however, no reason why the machine should always be used in such manner: there is nothing in its construction which obliges us to do so. It would be quite possible for the machine to try out variations of behaviour and accept or reject them ... and I have been hoping to make the machine do this ... This is possible because, without altering the design of the machine itself, it can, in theory at any rate, be used as a model of any other machine, by making it remember a suitable set of instructions.

(Turing, 1946)

And he also discussed why this would be possible using the ACE:

The ACE is, in fact, analogous to the 'universal machine' described in my paper on computable numbers. This theoretical possibility is attainable in practice, in all reasonable cases, at worst at the expense of operating slightly slower than a machine specially designed for the purpose in question. Thus, although the brain may in fact operate by changing its neuron circuits by the growth of axons and dendrites, we could nevertheless make a model, within the ACE, in which this possibility was allowed for, but in which the actual [hardware] construction of the ACE did not alter, but only the remembered data, describing the mode of behaviour applicable at any time.

(Turing, 1946)

Turing's target model, to be made 'within the ACE,' is based on the action of the brain, its alteration of neural circuits through the growth of axons and dendrites. His words to Ashby will not be appear exactly the same in later sources, and yet it will be possible to keep track of their evolution in the context of his slightly changing terminology. By that point in late 1946, Turing seems to have already understood that machine intelligence would require a new model of computation, one necessarily different from his disciplined universal machine presented in 'On Computable Numbers' (1936).

3.2.3   *February 1947: wanting a machine that can learn from experience*

Until the end of 1946, there seems to be no record of Turing using the word 'learning' explicitly. He began to do so in his lecture to the London Mathematical Society in February 1947, when he made a striking connection between between 'intelligence' and 'learning:'

One can imagine that after the machine had been operating for some time, the instructions would have altered out of all recognition ... In such a case one would have to admit that the progress of the machine had not been foreseen when its original instructions were put in. It would be like a pupil who had *learnt* much from his master, but had added much more by his own work. When this happens I feel that one is obliged to regard the machine as showing *intelligence* ... What we want is a machine that can *learn* from experience. The possibility of letting the machine alter its own instructions provides the mechanism for this.

(Turing, 1947, p. 393, my emphasis)

Here it becomes clear that Turing came to identify machine intelligence with machine learning.

Up to that point, Turing had concentrated his postwar work on the hardware architecture of the ACE and did not yet have a computational model to represent the mathematics of learning.[64]

A few months after his lecture, however, it was decided that Turing would take a sabbatical leave from the NPL. His plans and intentions were described in an official memorandum from the NPL director, Charles Darwin, dated July 23, 1947:

As you know Dr. A. Turing ... is the mathematician who has designed the theoretical part of our big computing engine ...
He wants to extend his work on the machine still further towards the biological side. I can best describe it by saying that hitherto the machine has been planned for work equivalent to that of the lower parts of the brain, and he wants to see how much a machine can do for the higher ones; for example, could a machine be made that could learn by experience?

(Darwin, 1947)

Darwin's memorandum confirms that Turing was determined to pursue the research plan he had told Ashby about. He wanted to find a mathematical model to make the machine learn from experience by imitating the 'higher' parts of the brain.

3.2.4   *Summer 1948: towards a mathematical model for machine learning*

Turing's 'Intelligent Machinery' report (1948) is a major source for his views on machine learning as the basis for machine intelligence. In contrast to his universal machine, whose logical structure is organized to imitate other machines under strict discipline, Turing introduced a new machine conception he called 'unorganized machines' (p. 416). As working model hypotheses, he presented networks of neuron-like Boolean elements that are initially randomly connected. Turing performed several experiments on paper to see if such a network of neurons could acquire certain specific topologies, associated with certain logical functions, in response to external perturbations.[65]

While the universal machine is completely disciplined, the unorganized machine should continuously learn and redefine its logical structure by a flexible combination of 'discipline' and 'initiative.' Turing distinguished the 'direct method' and the 'other method' for educating the unorganized machine (pp. 429–430). The former, similar to what he had imagined in 1936 for the universal machine, would 'take the form of programming the machine to do every kind of job that could be done.' Then '[b]it by bit,' he added, 'one would be able to allow the machine to make more and more "choices" or "decisions"' (p. 429). Eventually, it would be possible to program the machine 'to make its behaviour be the logical result of a comparatively small number of general principles.' Should these become sufficiently general, 'interference would no longer be necessary, and the machine would have "grown up."' This relates to what Turing metaphorically called 'screwdriver interference,' since it involves reprogramming and rebooting. The other method would rely more on 'paper interference,' dispensing with rebooting. In this case, the machine's education would be more centered on self-initiative and learning from experience.

It is worth quoting some of Turing's conceptual remarks along these lines, first in his 1948 section, §7, 'Education of machinery:'

> If we are trying to produce an intelligent machine, and are following the human model as closely as we can, we should begin with a machine with very little capacity to carry out elaborate operations or to react in a disciplined manner to orders (taking the form of interference). Then by applying appropriate interference, mimicking education, we should hope to modify the machine until it could be relied on to produce definite reactions to certain commands.
>
> (Turing, 1948, p. 422)

(Note that 'mimicking education' amounts to learning through education.) It is also worth reproducing here remarks from Turing's 1948 section, §12, 'Discipline and initiative,' which can reveal his nuanced views relating the universal machine to the unorganized machine:

> If the untrained infant's mind is to become an intelligent one, it must acquire both discipline and initiative ... To convert a brain or machine into a universal machine is the extremest form of discipline. Without

something of this kind one cannot set up proper communication. But discipline is certainly not enough in itself to produce intelligence. That which is required in addition we call initiative. This statement will have to serve as a definition. Our task is to discover the nature of this residue as it occurs in man, and to try and copy it in machines.

(Turing, 1948, p. 429)

Turing's concept of initiative is a challenging interpretive problem, as his 1948 NPL report is the only source where it appears. Chapter 7 presents a discussion that covers the influence of the Victorian novelist, Samuel Butler (1835–1902), on Turing, and may stimulate future research on Turing's concept of initiative. We do know that he conceived of initiative as related to *search*, writing that 'intellectual activity consists mainly of various kinds of search' (p. 431).

Turing summarized the contributions of his 1948 report as follows:

The possible ways in which machinery might be made to show intelligent behaviour are discussed. The analogy with the human brain is used as a guiding principle. It is pointed out that the potentialities of the human intelligence can only be realised if suitable education is provided. The investigation mainly centres round an analogous teaching process applied to machines. The idea of an unorganised machine is defined, and it is suggested that the infant human cortex is of this nature. Simple examples of such machines are given, and their education by means of rewards and punishments is discussed.

(Turing, 1948, pp. 431–432)

In such a short summary of his report, Turing twice refers to 'analogy' ('analogous'). This suggests that he did indeed conceive of machine intelligence as analogous to human intelligence, not identical to it.

His views on how to make an intelligent machine through a process of learning and education would be further elaborated in 1950.

3.2.5 1950: can learning machines be programmed to imitate the workings of a child's mind?

Turing devoted a large part (§7) and indeed the climax of the argument of his seminal paper, 'Computing Machinery and Intelligence' (1950) to describing his views on how to build 'learning machines'. Again, he described machine intelligence as the machine's ability to learn for itself from experience, and he emphasized that this was the proper response to Lady Lovelace's objection, which stated that '[t]he machine can do whatever we know how to order it to perform' (p. 450).

Also, in his 1950 text, Turing stated his aim to model and simulate a child's mind. This was in continuity with his 1948 reference to 'the untrained infant's mind.' In 1950, he wrote: 'Instead of trying to produce a programme to simulate the adult mind, why not rather try to produce one which simulates the child's?' (p. 455). Turing thus defined the problem of how to program a machine to be intelligent as the problem of finding appropriate programs (i) to imitate the workings of a child's mind and (ii) to adapt for the machine the educational process that a human child goes through. He also addressed a difficult conceptual problem in the very idea of machine learning:

> The idea of a learning machine may appear paradoxical to some readers. How can the rules of operation of the machine change? ... They should describe completely how the machine will react whatever its history might be, whatever changes it might undergo. The rules are thus quite time-invariant. This is quite true. The explanation of the paradox is that the rules which get changed in the learning process are of a rather less pretentious kind, claiming only an ephemeral validity.
>
> (Turing, 1950, p. 458)

This passage shows that, for Turing, machine learning should distinguish between hard and soft rules of behavior. In light of his constant reminder of that he was relying on an analogy with the human development of intelligence, that distinction was probably inspired by animal learning, where the most basic behavioral patterns of an individual of a given species are preserved.

3.2.6   c. 1951: the method of machine intelligence is learning from experience

Turing clearly argued that machine intelligence based on learning was not about satisfying some narrow metric. His approach to getting a machine to pass an intelligence test would involve its general education from experience, and passing a test should be a consequence of such a process, not a goal in itself, otherwise it would be a gross form of cheating:

> This process [of machine learning] could probably be hastened by a suitable selection of the experiences to which it was subjected. This might be called 'education'. But here we have to be careful. It would be quite easy to arrange the experiences in such a way that they automatically caused the structure of the machine to build up into a previously intended form, and this would obviously be a gross form of cheating, almost on a par with having a man inside the machine.
>
> (Turing, 1951b, p. 473)

That is, educating a machine to beat a certain performance metric or to impress an observer by demonstrating a certain skill is different from thinking. It should be considered a 'man inside the machine' stratagem. It is important to note that this passage shows that Turing *did* care about the method used by the machine.

3.2.7   *c. 1951: could a chess-playing machine learn from its experience?*

Turing reiterated his belief that a machine could be made to play chess well by learning from its experience in a chapter that he wrote in *c.* 1951 to appear in a book (1953).[66]

He began the text with the question: what does it mean to make a machine to play chess? Then he allowed several possible interpretations of this question, the fourth one being: 'Could one make a machine to play chess, and to improve its play, game by game, profiting from its experience?' (p. 569). His answer reflects the fact that his research was still in an initial stage: 'I should answer "I believe so. I know of no really convincing argument to support this belief and certainly of none to disprove it."'

His shifting of his research project to a mathematical theory of embryology, related to machine intelligence as part of his brain studies, and his tragic death some two and a half years later, would not allow much progress to be made.

3.2.8   *January 1952: learning how to learn*

Finally, Turing discussed machine learning at length at the BBC roundtable in January 1952. Perhaps his most significant remark concerned his expectations and hopes for the possibility of a machine *learning how to learn*. For Turing, such a 'snowball effect' would be an important indicator of whether research in machine learning was going in the right direction:

> *Turing*: ... One will have to find out how to make machines that will learn more quickly if there is to be any real success. One hopes too that there will be a sort of snowball effect. The more things the machine has learnt the easier it ought to be for it to learn others. In learning to do any particular thing it will probably also be learning to learn more efficiently. I am inclined to believe that when one has taught it to do certain things one will find that some other things which one had planned to teach it are happening without any special teaching being required. This certainly happens with an intelligent human mind, and if it doesn't happen when one is teaching a machine there is something lacking in the machine.
>
> (Turing et al., 1952, p. 497)

Note that Turing again explicitly identified the 'intelligent human mind' with the ability to learn in general and to learn how to learn.

Overall, this section has shown that Turing's identification of machine intelligence with machine learning spanned a chronological arc from late 1945 to early 1952, that is, from his earliest to his latest communications and writings on machine intelligence since the end of World War II. In light of the textual evidence presented here, it is possible to conclude that Turing could hardly have conceived of the possibility of attributing intelligence to a machine unless its intelligence was based on learning from experience.

3.3   In search of an empirical basis for think$_2$, 1948–1952

No matter how intelligent a machine may be, it turns out that the *natural order* (§1), the conventional wisdom that machines cannot think, may prevent an observer from attributing the property of thinking or intelligence to a machine. Therefore, to explain machine intelligence in terms of the method employed by the machine makes it unimpressive, 'whatever it may be, for it must be mechanical' (Turing, 1950, p. 450).

Aware of this problem, Turing seems to have realized that machine intelligence needed an empirical basis that could neutralize confirmation bias. Both because he used human intelligence (think$_1$) as an inspiration for constructing machine intelligence (think$_2$) and because his interlocutors could neither suspend their concept of intelligence nor generalize it beyond human intelligence, machine intelligence would have to be judged by the empirical standards associated with *human* intelligence. But since a machine such as a digital computer is not a human being, the best it can do is *imitate* a human being. A blind evaluation of the machine's imitation capabilities sets the stage.

This section examines Turing's remarks on the perception of intelligence in the context of the problem of establishing an empirical basis for machine intelligence (think$_2$).

3.3.1   *1948: The idea of 'intelligence' is itself emotional*

In his report 'Intelligent Machinery' (1948), the first two objections to the possibility of machine intelligence outlined by Turing were identified by (a) and (b):[67]

> The objections (a) and (b), being purely emotional, do not really need to be refuted. If one feels it necessary to refute them, there is little to be said that could hope to prevail, though the actual production of the machines would probably have some effect ... These arguments cannot be wholly

ignored, because the idea of 'intelligence' is itself emotional rather than mathematical.

(Turing, 1948, p. 411)

That is one of the passages that has been selected and emphasized in an earlier interpretation of Turing's concept of intelligence (Proudfoot, 2013). However, the textual context in which Turing observed the emotional aspect of intelligence in constructing his 1948 argument has been neglected and will now be restored.

The passage appears at the beginning of Turing's text, in dialogue with certain objections to the possibility of intelligent machines. Turing sharply observed that the *idea* of intelligence—that is, the conception, the way it is construed and perceived in culture—is emotional rather than mathematical. He then spent most of his 1948 argument developing how machine learning could be achieved through mathematical modeling. The order of the logical steps in his argument must be observed. It was not until later in his text, in the concluding section (§13) entitled 'Intelligence as an emotional concept,' that Turing returned to this view of intelligence. Despite all the efforts at mathematical modeling, he pointed out, it must be considered that the attribution of intelligence would still involve an emotional component, which is one of *two* components:

> The extent to which we regard something as behaving in an intelligent manner is determined as much by our own state of mind and training as by the properties of the object under consideration. If we are able to explain and predict its behaviour or if there seems to be little underlying plan, we have little temptation to imagine intelligence. With the same object therefore it is possible that one man would consider it as intelligent and another would not; the second man would have found out the rules of its behaviour.
>
> (Turing, 1948, p. 431)

Turing considers that the perception of intelligence depends on two elements: the subjective element ('our own state of mind and training' as external observers) and the objective element ('the properties of the object under consideration'). He suggests that these two elements are equally important (note the 'as much') for us to 'regard something as behaving in an intelligent manner.'

Immediately following the above excerpt, Turing described an 'idealized form of an experiment' he had conducted. The experiment was a preliminary version of his imitation game. In describing it, Turing emphasized how a human observer would *perceive* a chess-playing machine without knowing *a priori* which player was the human and which was the machine.

In summary, given the logical structure of Turing's 1948 text, I propose to interpret his goal in the above passage as emphasizing that success on the objective component—to which he devoted most of his attention and the core of his text—will not be sufficient for a machine to be granted the property of being intelligent by an external observer. The machine must also *appear* intelligent to the observer, taking into account the observer's own state of mind and training. Of course, if the observer is conditioned to think of humans and only humans as intelligent, then this means that the machine must act, lie, impersonate, and imitate a human, which it is *not*. In other words, the machine must use think$_2$ to imitate think$_1$.[68] If it succeeds, then it deserves to be called intelligent, and think$_2$ would eventually permeate everyday life and common sense and deserve to be called thinking.

3.3.2   *1950: explaining mechanical intelligence makes it unimpressive*

Turing would return to his view of intelligence as an emotional concept in his 1950 paper, when discussing the fifth objection, 'Arguments from various disabilities.' It resembles the 1948 excerpt above:

> Usually if one maintains that a machine can do one of these things and describes the kind of method that the machine could use, one will not make much of an impression. It is thought that the method (whatever it may be, for it must be mechanical) is really rather base. Compare the parenthesis in Jefferson's statement quoted on pp. [445–446, namely '(and not merely artificially signal, an easy contrivance)'].
> (Turing, 1950, pp. 449–450)

Again, Turing emphasizes the emotional character of the *idea* of intelligence in the context of his discussion of objections, i.e., in dialectic with the reaction of his interlocutors, which he considered biased against the machines.

3.3.3   *1951: 'learning by experience' would be impressive*

In *c*. 1951, Turing noted that if a machine is capable of learning by experience, then its ability to imitate human intelligence, which is required to be perceived as intelligent, 'would be much more impressive:'

> It is clearly possible to produce a machine which would give a very good account of itself for any range of tests, if the machine were made sufficiently elaborate. However, this again would hardly be considered an adequate proof. Such a machine would give itself away by making the same sort of mistake over and over again, and being quite unable to correct itself, or to be corrected by argument from outside. If the machine

were able in some way to 'learn by experience,' it would be much more impressive.

(Turing, 1951b, p. 473)

Turing resumed: 'As I see it, this education process would in practice be an essential to the production of a reasonably intelligent machine within a reasonably short space of time' (p. 473). 'The human analogy alone,' he concluded, 'suggests this.'

It is worth noting in these passages how Turing relates the machine's ability to learn to its performance in imitating human intelligence. We have seen (Section 3.2) that he was thinking in terms of an analogy with the human cognitive development in infancy.

3.3.4   *1952: thinking as mental processes that we don't understand*

In a discussion with Richard Braithwaite (1900–1990), also in the presence of Max Newman and Geoffrey Jefferson, on the BBC roundtable in January 1952, Turing would reiterate the point that explaining machine intelligence makes it unimpressive:

> *Braithwaite*: But could a machine really do this [learn by analogy]? How would it do it?
> *Turing*: I've certainly left a great deal to the imagination. If I had given a longer explanation I might have made it seem more certain that what I was describing was feasible, but you would probably feel rather uneasy about it all, and you'd probably exclaim impatiently, 'Well, yes, I see that a machine could do all that, but I wouldn't call it thinking.' As soon as one can see the cause and effect working themselves out in the brain, one regards it as not being thinking, but a sort of unimaginative donkey-work. From this point of view, one might be tempted to define thinking as consisting of 'those mental processes that we don't understand'. If this is right, then to make a thinking machine is to make one which does interesting things without our really understanding quite how it is done.
> (Turing et al., 1952, p. 500)

On the one hand, this shows that Turing had consolidated the view, not at all obvious in the early 1950s, which he had already expressed in 1948 and in 1950, that the idea or conception of intelligence is emotional. In light of this, the ability to meet the standard expected by an observer, in the case of a (biased) human observer, *human* intelligence (think$_1$), is indeed a key ingredient for an agent to be credited with intelligence.

On the other hand, as Turing also noted, this is only one 'point of view,' and it is a negative aspect of the phenomenon of intelligence.

If it were not complemented by a positive theory, intelligence would become an elusive concept to be established by convention alone. The idea that Turing would have conceived of intelligence in this way, which is useless for his main project of building an intelligent machine, is nonsense.

The two objections ('a' and 'b') formulated by Turing in 1948 and considered 'purely emotional' may suggest that he was aware of how important the concept of intelligence was from a social, cultural, and even psychological point of view. It seems that he proposed the imitation of human intelligence (think$_1$) as the empirical basis for machine intelligence as a way to neutralize the confirmation bias towards the conservative position that machines have never thought, cannot think, and will never think.

3.3.5   1952: confirmation bias against think$_2$ and the question of the method of machine intelligence revisited

Finally, also at the BBC roundtable in 1952, in the presence of Turing, Newman had a few exchanges with Jefferson that are worth considering.

First, in his own words and in the presence of Turing, Newman added to Turing's point about the confirmation bias against machine intelligence:

> *Jefferson*: ... I don't see how a machine could as it were say 'Now Professor Newman or Mr. Turing, I don't like this programme at all that you've just put into me, in fact I'm not going to have anything to do with it.'
>
> *Newman*: One difficulty about answering that is one that Turing has already mentioned. If someone says, 'Could a machine do this, e.g. could it say "I don't like the programme you have just put into me", and a programme for doing that very thing is duly produced, it is apt to have an artificial and ad hoc air, and appear to be more of a trick than a serious answer to the question. It is like those passages in the Bible, which worried me as a small boy, that say that such and such was done 'that the prophecy might be fulfilled which says' so and so. This always seemed to me a most unfair way of making sure that the prophecy came true. If I answer your question, Jefferson, by making a routine which simply caused the machine to say just the words 'Newman and Turing, I don't like your programme', you would certainly feel this was a rather childish trick and not the answer to what you really wanted to know. But yet it's hard to pin down what you want.
>
> (Turing et al., 1952, pp. 501–502)

Demands such as Jefferson's, Newman seems to suggest, could create *ad hoc* and perhaps biased barriers to the discussion.

Now, in connection with Turing's test, Newman said something that may sound contradictory to Turing's prohibition of cheating or trickery quoted above:

> *Newman:* ... Of course, this is a dull, plodding method [brute-force table lookup], and you could improve on it by using a more complicated routine, but if I have understood Turing's test properly, you are not allowed to go behind the scenes and criticise the method, but must abide by the scoring on correct answers, found reasonably quickly.
>
> (Turing et al., 1952, p. 496)

Newman's understanding of Turing's test can be seen as supporting its interpretation as nothing but a social convention designed to evaluate an emotional concept. However, Newman himself used the caveat: 'if I have understood Turing's test properly.' There is no record of Turing's response to that, and the assumption that Turing should have intervened and contradicted Newman, adding complexity to an already complicated discussion with Jefferson, seems unreasonable. In any case, given Turing's direct comment in *c.* 1951 (Section 3.2) on the method and the real purpose of preparing a machine for an intelligence test — i.e., to test for true machine learning —, Newman's comment can be regarded as secondary.

The conceptual chronologies presented in these last two sections can pave the way for a more nuanced view of Turing's proposed theoretical foundation and empirical basis for machine intelligence.

3.4   Imitation of think$_1$ is the empirical basis for think$_2$

This section examines Turing's use of the concept of imitation as it appears in primary sources. It supports the interpretation that Turing conceived of imitation, particularly of think$_1$, as the empirical basis for the development of machine intelligence (think$_2$).

3.4.1   *Turing's concept of imitation was technical*

It is worth starting by narrowing down the meaning of Turing's use of 'imitation.' In his lecture to the London Mathematical Society in February 1947, he explained the possibility of the ACE, as a realization of his conceptual universal machine (1936), to imitate other machines:

> Let us now return to the analogy of the theoretical computing machines with an infinite tape. It can be shown that a single special machine of that type can be made to do the work of all. It could in fact be made to work as a model of any other machine. The special machine may be

called the universal machine; it works in the following quite simple manner. When we have decided what machine we wish to *imitate* we punch a description of it on the tape of the universal machine. This description explains what the machine would do in every configuration in which it might find itself. The universal machine has only to keep looking at this description in order to find out what it should do at each stage. Thus the complexity of the machine to be *imitated* is concentrated in the tape and does not appear in the universal machine proper in any way.

(Turing, 1947, p. 383, my emphasis)

This passage gives a first hint that Turing's concept of imitation was technical.

Later in (1951a), Turing again explained his concept of imitation in connection with the universality property, but now explicitly discussing machine intelligence:

A digital computer is a universal machine in the sense that it can be made to replace any machine of a certain very wide class. It will not replace a bulldozer or a steam-engine or a telescope, but it will replace ... any machine into which one can feed data and which will later print out results. In order to arrange for our computer to *imitate* a given machine it is only necessary to programme the computer to calculate what the machine in question would do under given circumstances, and in particular what answers it would print out. The computer can then be made to print out the same answers. If now some particular machine can be described as a brain we have only to programme our digital computer to *imitate* it and it will also be a brain. If it is accepted that real brains, as found in animals, and in particular in men, are a sort of machine it will follow that our digital computer, suitably programmed, will behave like a brain.

(Turing, 1951a, pp. 482–483, my emphasis)

Note Turing's focus on imitating input-output behavior, while implicitly assuming a certain abstraction of the internal structure of the imitating and imitated machines. He went on to acknowledge that this 'involves several assumptions which can quite reasonably be challenged.' In particular, 'the machine to be imitated must be more like a calculator than a bulldozer.' He considered machines for intellectual activity such as a calculator to be 'mechanical analogues of brains' (p. 483).

Now, an archival source, also from mid-1951, and new to Turing scholarship as far as I know, may give more contour and provide some further insight into Turing's views. He wrote to Beatrice Worsley (1921–1972), whose Ph.D. thesis at Cambridge University he helped to supervise (my emphasis):

Dear Miss Worsley,

I was interested in your work on the relation between computers and Turing machines. I think it would be better though if you could try and find a realtion [*sic*] between T machines and infinite computers, rahter [*sic*] than between finite T machines and computers. The relation that you suggest is rather too trivial. The fact is that the motions of either a finite T machine or a finite computer are ultimately periodic, and therefore any sequence computed by them is ultimately periodic. It is easy therefore in theory to make one *imitate* the other, though the size of the *imitating* machine will (if this technique is adopted) have to be of the order of the exponential of the size of the *imitated* machine. Probably your methods could prove that this exponential relation could be reduced to a multiplicative factor.

Yours sincerely,

A. M. Turing[69]

An in-depth analysis of Turing's remarks in this letter from the perspective of his views on machine intelligence is due in future work. For the purposes of this section, it is worth noting that Turing's references to 'imitation' in discussing the possibility of machine intelligence suggest faithfulness and depth in relation to human intelligence, not deception and shallowness. But especially since 1948, aware of the confirmation bias against the perception of machine intelligence, he had to shift from chess-based imitation to conversation-based imitation, since only the latter, a hallmark of human intelligence, would be accepted as 'intelligence.'[70] With this move, faithful imitation must become deception. In an unrestricted conversation, the machine could be directly questioned about its nature and give itself away, unless it resorted to acting, lying, and deception.

3.4.2 *The reconstruction of think₁ is an intractable problem*

The surviving minutes of the October 1949 seminar (Turing et al., 1949), 'The Mind and the Computing Machine,' at the Philosophy Department of the University of Manchester,[71] captured a crucial exchange between Turing and Michael Polanyi on the possibility of specifying the human mind:

> POLANYI: interprets this [an observation by Turing on machine consciousness] as suggestion that the semantic function can ultimately be specified; whereas in point of fact a machine is fully specifiable, while a mind is not.

TURING: replies that the mind is only said to be unspecifiable because it has *not yet been* specified; but it is a fact that it would be impossible to find the programme inserted into quite a simple machine — and we are in the same position as regards the brain. The conclusion that the mind is unspecifiable does not follow.

POLANYI: says that this should mean that you cannot decide logical problems by empirical methods.

(Turing et al., 1949, his emphasis)

Turing compares the human mind to a computer program. The point of the comparison is to neutralize an extraneous aspect of the dissimilarity between the two, namely that we tend to think of the inner logic of the computer program as simple because we have designed it ourselves, whereas the inner logic of the human mind is still somewhat mysterious to us.[72] We may suppose that Polanyi's comment — 'you cannot decide logical problems by empirical methods' — would have immediately reminded Turing of his codebreaking experience at Bletchley Park during World War II. Polanyi was saying that to the face of an extraordinary scientist whose wartime task was to learn the inner logic of the Enigma machine empirically by observing its outer behavior alone, but the prevailing state secrecy would not allow Turing to refer to that experience. The passage shows that Turing rejects the theoretical impossibility of specifying the logic (the 'program') of the human mind ($think_1$), while acknowledging the intractability of reverse engineering it. This is true even for a computer program, and it is obviously much worse for the human mind.

Turing seems to suggest in his discussion with Polanyi that the human mind is too complex, and that the best scientific research strategy must therefore be a hypothetico-deductive one, i.e. guessing directly at $think_2$; taking inspiration from $think_1$, but not trying to reconstruct it exactly. Note that this interpretation is consistent with the view that Turing proposed a theoretical basis for machine intelligence centered on learning (Section 3.2).

That Turing did not try to reconstruct $think_1$ in its nature can also be seen from one of his interventions at the BBC roundtable in January 1952. Jefferson was asked to open the discussion, and he ended his first comment with a question to Turing:

Jefferson: Turing, what do you think about it? Have you a mechanical definition?

Turing: I don't want to give a definition of thinking, but if I had to I should probably be unable to say anything more about it than that it was a sort of buzzing that went on inside my head. But I don't really see that we need to agree on a definition at all. The important thing is to try to draw a line between the properties of a brain, or of a man, that we want to discuss, and those that we don't ... I would like to suggest a particular kind of *test* that one might apply to a machine.

(Turing et al., 1952, pp. 494–495, Turing's emphasis)

Turing concludes his refusal to give a definition of *human* thinking by referring to his proposed test for *machine* intelligence. Again, this suggests that the constructivist view of Turing's tests, which argues that Turing intended them as tests for the human judges, is false.

3.4.3 Think$_1$ and think$_2$ are intrinsically different

Further conceptual clarity can be achieved by examining how Turing understood the physical and architectural differences between the digital computer and human brain.

The Turing-Jefferson exchanges, discussed in more detail in Chapters 4 and 5, provide an excellent window to appreciate Turing's distinction between think$_1$ and think$_2$. For a brief contextualization, in his Lister Oration, delivered and published in June 1949, Jefferson (1949a) criticized 'the physicists and mathematicians,' meaning scientists associated with cybernetics, who he thought proposed an identity between machine and brain:

We feel perhaps that we are being pushed, gently not roughly pushed, to accept the great likeness between the actions of electronic machines and those of the nervous system.

(Jefferson, 1949a, p. 1105)

Jefferson absolved engineers who, in his view, did not abuse the analogy between the computer and the brain:

To be just, nothing more than analogy is claimed by most of their constructors (some, like Professor Williams, do not go so far even as that), but there is a grave danger that those not so well informed will go to great lengths of fantasy.

(Jefferson, 1949a, p. 1108)

Jefferson thought that cyberneticians failed to observe that the nervous impulse is a continuous signal, and further it is not really an electrical but an electro-chemical phenomenon. This, for Jefferson, was prohibitive of a strong analogy or an identity with computing machines and the phenomena of electronic information storage and processing (pp. 1107–1108).

Jefferson argued for a sort of incommensurability between the nervous system and computing machines. He devoted a full section of his Lister

Oration to discuss four topics relative to the nervous impulse (in my words): (i) speeds of the electronic computing machine compared to the nervous impulse, (ii) speed of thought, (iii) counter-evidence to the electrical nature of the nerve impulse, and (iv) methods of study and the chemical agencies of the nervous system. Later, at the BBC roundtable in January 1952, he summed it up: 'Man is essentially a chemical machine' (1952, p. 502).

Turing addressed this aspect of Jefferson's criticism in his formulation of 'The argument from continuity in the nervous system' (1950, pp. 451–452). 'The nervous system,' Turing admitted, 'is certainly not a discrete-state machine,' for '[a] small error in the information about the size of a nervous impulse impinging on a neuron, may make a large difference to the size of the outgoing impulse' (1950, p. 451). However, he argued, this does not mean that the behavior of the nervous system cannot be imitated by a discrete-state system. He argued that the design of the imitation game neutralizes this physical, architectural, structural, ontological difference.[73]

Less than a year later (1951a), Turing referred to his 'main problem' as that of 'how to programme a machine to imitate a brain, or as we may say more briefly, if less accurately, to think' (p. 485). It is worth noting his explicit reference to the use of the word, 'think,' as less accurate. But we have also seen another aspect that he explicitly noted, namely that the status of the mechanism underlying the intellectual activity of the digital computer is *necessarily* not identical to that of the human brain. In short, Turing's argument can best be interpreted as distinguishing between the mechanism of the machine (think$_2$) and that of the human (think$_1$), even though the intelligence they produce might correspond at the behavioral level.

3.4.4   The challenges of constructing think$_2$ to imitate think$_1$

Also in his BBC broadcast in May 1951, Turing said: 'it is not altogether unreasonable to describe digital computers as brains.' If suitably programmed, he added, they 'will behave *like* a brain' (1951a, p. 482, my emphasis).

In that lecture, he gave what is probably his most direct articulation of the challenge of constructing an appropriate think$_2$ in a machine. He considered two hardware technology requirements: sufficient speed and storage capacity. He promptly dismissed the speed issue, believing that existing digital computing technology was good enough to imitate the human brain. But he considered storage to be a serious bottleneck at the time:[74]

> Our present computers probably have not got the necessary storage capacity, though they may well have the speed. This means in effect that if we wish to imitate anything so complicated as the human brain we

need a very much larger machine than any of the computers at present available.

(Turing, 1951a, p. 483)

It is worth noting that, for Turing, *all* that was still needed in terms of hardware was storage capacity:

It should be noticed that there is no need for there to be any increase in the complexity of the computers used. If we try to imitate ever more complicated machines or brains we must use larger and larger computers to do it. We do not need to use successively more complicated ones. This may appear paradoxical, but the explanation is not difficult. The imitation of a machine by a computer requires not only that we should have made the computer, but that we should have programmed it appropriately. The more complicated the machine to be imitated the more complicated must the programme be.

(Turing, 1951a, p. 483)

This was quite an assertive statement on what can be seen as an important question of computer architecture regarding the possibility of imitating the human brain, and which may still be considered an open question today.[75] It is consistent with his hardware design choices for the ACE in the mid-1940s, namely to shift complexity out of hardware and into software (cf. Carpenter and Doran, 1977).

Turing went on to state what he considered to be the decisive factor in answering his question, whether machines can think, in the affirmative:

[I]t seems that the wisest ground on which to criticise the description of digital computers as 'mechanical brains' or 'electronic brains' is that, although they might be programmed to behave *like* brains, we do not at present know how this should be done. With this outlook I am in full agreement. It leaves open the question as to whether we will or will not eventually succeed in finding such a programme. I, personally, am inclined to believe that such a programme will be found.

(Turing, 1951a, p. 484, my emphasis)

Turing admits his uncertainty of whether a program would be found to program a machine to 'behave like brains' or, as it has been suggested in this chapter, to construct think$_2$ to imitate think$_1$. And yet he clearly states his belief that such a program would be found.

Another archival source, another letter from Turing to Worsley from mid-1951, also new to Turing scholarship as far as I know, may give more contour and provide some further insight into Turing's views. He wrote (his emphasis):

> I think any attempt to draw any *sharp* line between what machine and brain can do will fail. I think it is largely a quantitative matter. Probably one needs immensely more storage capacity then [*sic.*] we have got, and possibly more than we shall ever have. Perhaps we may have enough capacity, but just won't find an appropriate programme. Naturally one won't make a man that way ever. It'll just be another species of the thinking genus.[76]

Turing reinstates that the complexity of the problem of constructing $think_2$ was largely a matter of storage capacity and of finding an appropriate program. While storage capacity was an engineering problem, finding an appropriate program ($think_2$) to imitate the mind ($think_1$) was a scientific problem. But Turing's first and last two sentences are striking. On the one hand, the first sentence clearly states his view that ultimately there can be no clear boundary separating what machine and brain can do at the behavioral level of intellectual activity. On the other hand, the last two sentences suggest that he really did think that the inner process of the intelligent machine species belonged to a different kind of thinking, which we may indeed call $think_2$.

3.5   How mechanical can $think_1$ be?

In 1950, Turing made this assertive statement:

> The reader must accept it as a fact that digital computers can be constructed, and indeed have been constructed, according to the principles we have described, and that they can in fact mimic the actions of a *human computer* very closely.
>
> (Turing, 1950, p. 438, my emphasis)

Note that, from an epistemological perspective, he views human computing as a proof of concept for his imitation-based approach. In 1951, he made another statement whose difference from the earlier one is worth emphasizing:

> My contention is that machines can be constructed which will simulate the behaviour of the human mind very closely.
>
> (Turing, 1951b, p. 472)

Turing moves from the imitation of the intellectual activity of a human computer, taken as a matter of proven fact, to the imitation of the intellectual activity of the human mind as a whole, taken as a kind of conjecture.[77]

In fact, while he did not attempt to literally reconstruct the human mind, he did attempt to explore the extension and limits of mechanism as constitutive of the human mind. Already in 1950, Turing had given this clear view on the question of the extension and limits of a mechanical mind as part of the real human mind. He used the analogy of the 'skin of an onion':

> The 'skin of an onion' analogy is also helpful. In considering the functions of the mind or the brain we find certain operations which we can explain in purely mechanical terms. This we say does not correspond to the real mind: it is a sort of skin which we must strip off if we are to find the real mind. But then in what remains we find a further skin to be stripped off, and so on. Proceeding in this way do we ever come to the 'real' mind, or do we eventually come to the skin which has nothing in it? In the latter case the whole mind is mechanical. (It would not be a discrete-state machine however. We have discussed this.)
>
> (Turing, 1950, pp. 454–455)

The status of the mechanical mind, Turing suggested, might turn out to be either a proper part of, or equivalent to, the real mind. He saw the existence of a remaining core of the human mind, which might ultimately never be imitated by a machine, as an open empirical problem.

This passage reveals a key aspect of Turing's scientific philosophy. Although he did not conflate analogy with identity, being always clear about the abstract nature of his studies, he did seem to expect that his analogy between mind and machine would work in both directions. On the one hand, think$_2$ would be constructed to imitate think$_1$ and by drawing inspiration from it. On the other hand, Turing reasoned, the more of the intellectual activity of the human mind is successfully imitated by the machine, the greater will be the relative proportion of the mechanical mind as a constitutive part of the 'real' mind. And he was curious to see if there was any intellectual behavior that would ultimately remain beyond the reach of machines. If none, then there could be no distinguished real mind left — the whole mind would be mechanical.

At the end of the BBC roundtable in January 1952, Jefferson expressed his frustration not to find the instrumentalism he wanted to see in Turing:

> *Newman*: The logical plan of all of them [all digital computers] is rather similar, but certainly their anatomy, and I suppose you could say their physiology, varies a lot.

Jefferson: Yes, that's what I imagined — we cannot then assume that any one of these electronic machines is a replica of part of a man's brain even though the result of its actions has to be conceded as thought. The real value of the machine to you is its end results, its performance, rather than that its plan reveals to us a model of our brains and nerves ... It would be fun some day, Turing, to listen to a discussion, say on the Fourth Programme, between two machines on why human beings think that they think!

(Turing et al., 1952, pp. 505–506)

Contrary to Jefferson's wishes, Turing was anything but an instrumentalist. He distinguished the sense in which the machine could be said to think. But he did expect the use of the word to shift and eventually merge in everyday life and common sense in the presence of talking machines, he did expect his science to inform our interpretation of the human mind.

3.6   Conclusion

Turing's proposal of his imitation game and test has largely been understood as a rejection of the distinction between the imitation of intelligence and intelligence itself, or as a conflation of analogy and identity. A more nuanced view of Turing's philosophy of science for machine intelligence has been presented.

The proposed interpretation of Turing's scientific philosophy has been supported by conceptual chronologies of primary sources. The first chronology, 1945–1952, showed that Turing based his concept of machine intelligence in his working mathematical theory of machine learning, inspired by the human development of a child's brain in its social environment (Section 3.2). The second chronology, 1948–1952, clarified why and in what terms Turing sought for an empirical basis for machine intelligence, an approach that could neutralize confirmation bias against the machine (Section 3.3). Turing's concept of imitation was technical and overarching in his scientific philosophy. The epistemological importance of imitation, and in particular the establishment of the imitation of human intelligence as the empirical basis for machine intelligence, has been emphasized (Section 3.4).

The following is a summary of this proposed interpretation of Turing's scientific philosophy for machine intelligence. As realizations of his universal machine, digital computers were, for Turing, pieces of hardware capable of following instructions but not of thinking. However, a learning machine could be built as software on top of such hardware — just as the brain itself can be seen as software at the cellular (neural) level — and thus change its software structure based on experience. Turing drew inspiration from how the brain changes its neural circuits by growing and redistributing axons and dendrites. He noted critical differences between the computer and the brain, most notably that the digital computer is a discrete-state machine and thus cannot be said to be identical to the human brain (and nervous system), which is a continuous-state machine. But the learning machine could be made to imitate the brain in a general intelligence task such as conversation to the point of indistinguishability. Should this empirical achievement be realized, it could be taken as evidence that a significant part of the human intellectual activity can be understood as 'mechanical' by analogy, with the caveat that this term has a broader meaning for Turing than being computable by a Turing machine.

Turing stated his belief that a suitable computer program for imitating human intelligence would eventually be found. I have suggested that he was not an instrumentalist. He pursued his research to inform our understanding of both nature and society. While he did not aim to reconstruct the human mind literally, he did aim to explore by analogy — and to the limit — the extension of mechanism as constitutive of the human mind, i.e., how mechanical the mind could be.

4   The Controversy that Led to the Turing Test

Turing's 1950 paper is often considered a complex and multilayered text, and key questions about it remain largely unanswered. Why did Turing argue for learning from experience as the best approach to achieving machine intelligence? Why did he spend several years working with chess playing as a task to illustrate and test for machine intelligence, only to trade it out for conversation in 1950? Why did he refer to gender imitation in a test for machine intelligence? This chapter addresses these questions by noting a historical fact that has received little attention in the secondary literature, namely that Turing proposed his test in the context of a specific controversy over the cognitive capabilities of digital computers, most notably with physicist and computer pioneer Douglas Hartree, chemist and philosopher Michael Polanyi, and neurosurgeon Geoffrey Jefferson. Seen in its context, Turing's 1950 paper can be understood as a response to challenges posed to him by them. Turing did propose gender learning and imitation as one of his various imitation tests for machine intelligence, and this chapter argues that this was an important element in responding to Jefferson's suggestion that gendered behavior is causally related to the physiology of sex hormones.[78]

4.1   The purpose of the Turing test

Gandy's anecdote about Turing's motivation to write his 1950 paper (Chapter 1) has been previously mentioned in passing by Jack Copeland (2004, p. 433) and by Margaret Boden (2006, p. 1351), but has not yet been considered or analyzed.[79] It diverges from the common view of Turing's paper as proposing a crucial experiment for machine intelligence, or as it is often said, artificial intelligence (AI). But more than that, it suggests that Turing was engaged in a dialogue with 'philosophers and mathematicians and scientists' about the cognitive capabilities of digital computers. But what debate was that? Who were the interlocutors that Gandy suggests Turing sought to persuade? While the Turing test is widely known in philosophy, there have been no detailed historical studies of the origins and context of Turing's 1950 paper, beyond a few remarks made by biographers. I will assemble a mass of available sources which, although mostly known to

DOI: 10.4324/9781003300267-4

Turing scholars, have not yet been linked together and properly appreciated for the role they can play in sharpening our interpretation of Turing's 1950 proposal.

Seventy years have passed since Turing's famous 1950 proposal of an imitation game and test for machine intelligence, and its interpretations still vary widely. Chapter 2 gives an overview and brief discussion of representative classes of the interpretations of Turing's test. My goal in this chapter is not to discuss them further, but to position them in dialogue with Gandy's anecdote and the purpose of the Turing test.

First, interpreters disagree about whether Turing proposed a practical experiment to determine machine intelligence. On the one hand, philosophers such as Daniel Dennett (1984), Jack Copeland (2000b), Gualtiero Piccinini (2000), and James Moor (2001) have all provided support for viewing Turing's test as such an experiment. Dennett (1984) wrote that 'the Turing test, conceived as he conceived it, is (as he thought) plenty strong enough as a test of thinking' and added, 'I defy anyone to improve upon it' (p. 297). These interpretations have seen Turing's paper (1950) as an epic test for machines, seen as a species, as opposed to the human species. These readings have barely acknowledged the presence of the gendered variants of the imitation game that *are part of Turing's text* (pp. 442, 433–444). They read 'man' in Turing's *machine-imitates-man* variant as a masculine generic and generally argue that if one considers other primary sources and not just Turing's 1950 text, one must concede that Turing proposed not a gender test but a species test. One of my contentions is that Turing did propose gender learning and imitation as one of his various tests for machine intelligence. But those interpreters emphasize a *machine-imitates-human* test, which they read from Turing's text and think is the best experiment to decide whether machines have acquired human-level intelligence. Turing's text is read on the assumption that its purpose was to propose a practical experiment.

Dennett is an exception, as he later reconsidered and adopted a more deflationary position. After his experience with the Loebner Prize competition, which he described in (1997) as 'a fascinating *social* experiment' (my emphasis). He then wrote that the Turing test 'requires too much Disney and not enough science' and observed: 'The Turing Test is too difficult for the real world.' On the other hand, scientists such as Hayes and Ford (1995) and Drew McDermott (2014), although less certain about what Turing was trying to do in his paper, tried to take his 1950 proposal seriously as a practical experiment for machine intelligence, but found no scientific substance in it. Further, following their assumption that Turing had proposed a crucial experiment for AI, this class of readers complained about the quality of Turing's experimental description and design, be it for a gender or a species test. McDermott wrote: 'Considering the importance Turing's Imitation Game has assumed in the philosophy-of-mind literature of the last 50 years, it is a pity he was not clearer about what the game was exactly.'

Furthermore, there are those who have always suggested a deflationary view of Turing's 1950 proposal, if not denying altogether that Turing's test addresses an empirical question. Whitby (1996) acknowledged the role that Turing's proposal played in the early 1950s to inspire or drive research. However, he claimed that it soon became a distraction because Turing's test measures the human reaction to the machine, which is not a problem in artificial intelligence research. Like Gandy in his anecdote, as we saw in Chapter 1, Marvin Minsky said: 'the Turing test is a joke, sort of;' Minsky believed that Turing suggested his test 'as one way to evaluate a machine but he had never intended it as the way to decide whether a machine was really intelligent.'[80] And as we saw in Chapter 2, Noam Chomsky (1995) wrote that the question 'Can machines think?' 'is not a question of fact,' but one of language, and that Turing himself remarked that the question is 'too meaningless to deserve discussion' (p. 9).

Saygin et al. (2000) contributed a comprehensive survey of the 'Turing test' in the philosophy of mind and AI literature, which does not seek historical or philosophical depth.

Given the sheer heterogeneity of the secondary literature on the Turing test, one can conclude that there is still much room for the study of Turing's 1950 proposal from a more rigorous and historical perspective.

4.2   Core events of 1949, the crucial year

In June 1949, the computer pioneer, distinguished physicist, and then University of Cambridge professor Douglas Hartree published his *Calculating Instruments and Machines* (1949). He argued that the new electronic computing machines could do a lot, but should still be seen as nothing more than calculating machines (Section 4.4). On June 9, the eminent neurosurgeon and then professor at the University of Manchester, Geoffrey Jefferson, had delivered his Lister Oration (1949a) in London along the same lines, but going further. He outlined stringent requirements that would have to be met before a 'machine equals brain' thesis could be accepted. Jefferson argued that a machine should be able to write a sonnet and feel its meaning in order to be considered capable of thinking. Jefferson's Lister Oration was covered by *The Times*; the next day, when asked by a reporter for a response, Turing made a sharp and witty rebuttal, which also appeared in *The Times* (Section 4.6). Jefferson would have an actual influence on Turing's views after the December 1949 edition of a seminar entitled 'Mind and Computing Machine' in the Department of Philosophy at their university. The first edition had been held on October 27, 1949. These two seminars were co-chaired by the eminent chemist and Professor of Social

Studies at Manchester University, Michael Polanyi, who was also involved in the mind-machine controversy with Turing (Section 4.5). These three thinkers, all of whom at the time held fellowships of the Royal Society and university positions more prestigious than that of Turing, who was then a reader in the Department of Mathematics at the University of Manchester, attempted to impose limits on Turing's views on machine intelligence. From June to December 1949, as we shall see, Turing considered their challenges and responded to them in his famous 1950 paper. While these events are largely known to Turing scholars, their significance for understanding Turing's test proposal has largely been neglected.

4.3   The controversy in England, 1946–1950

Turing's dialogue with Hartree can be traced by direct citations from 1950 to 1951, which dealt with the possibility of learning machines.[81] To my knowledge, however, the influence of Hartree's opposition on Turing has received little attention. Andrew Hodges (2009) discussed Turing's test and argued for the Turing-Polanyi connection, writing that Polanyi 'encouraged him to publish his views' (p. 13). In his great Turing biography, Hodges (1983) provided several valuable primary and secondary sources. However, Hodges seems to struggle to understand Turing's text in Turing's own terms. For example, he writes that Turing's 'sexual guessing game' was 'in fact a red herring, and one of the few passages of the paper that was not expressed with perfect lucidity' (p. 415). Jonathan Swinton's *Alan Turing Manchester* (2019) provided many valuable new sources. Regarding Turing's test, Swinton emphasizes the Turing-Jefferson connection, noting that 'it was Jefferson's obtuseness that provoked Turing into developing this vivid image [the Turing test]' (p. 93). Swinton did not elaborate on that. Diane Proudfoot 2013 provided an interpretation of Turing's 1950 proposal that focused on a specific aspect of Turing's concept of intelligence, namely, Turing's 1948 observation that 'the idea of "intelligence" is itself emotional' (p. 411). More generally, Proudfoot concurred with Copeland's defense of the test as a sort of crucial experiment for artificial intelligence (2000b). I will also refer later to Darren Abramson (2011) and his location of material evidence that Turing read Jefferson's citations of René Descartes in his Lister Oration (1949a). My contention here is that we can connect specific findings such as this to build a more comprehensive, overarching interpretation of Turing's 1950 test that is historically grounded and also helps to explain some of the tensions in existing interpretations. This gap in the secondary literature can be illustrated by the general obliviousness to Gandy's anecdote, but also by key exegetical and historical questions that still seem to be largely unanswered:

- Why did Turing choose learning from experience as the best approach to achieving machine intelligence? In what context did he make this choice? Can we trace its intellectual history?
- If Turing was decided that conversation was the best intellectual task to illustrate, develop, and test machine intelligence, why did he work for several years from his war service in 1941 to his 1950 paper with chess playing as such a task and even reconsider it at the end of his 1950 paper?
- If Turing really favored a 'species' test for machine intelligence over a gender test in his 1950 paper, why did he refer so forcefully to gender imitation right at the beginning of the same exegetical unit?

Developing an account of key events in 1949 in the following sections enables me to offer relatively succinct answers to these questions. A more detailed account can be found elsewhere (Gonçalves, 2021). I suggest that Turing's test proposal (1950) can be best understood against the backdrop of the mind-machine controversy, particularly involving Hartree, Polanyi, and Jefferson in England from 1946 to 1950. Ultimately, this is the debate that led Turing to propose his famous imitation game and test for machine intelligence, and it explains Robin Gandy's anecdote.

4.4   Turing's exchanges with Douglas Hartree

Douglas Hartree (1897–1958), FRS since 1932 and Plummer Professor of Mathematical Physics and member of the Cavendish Laboratory at the University of Cambridge since October 1, 1946 (Darwin, 1958), had given a 'short series of lectures' at the University of Illinois in the early fall of 1948, which were published as *Calculating Instruments and Machines* circa June 1949. In the preface (p. v), dated May 1949, Hartree cited the Manchester 'Baby' computer, mentioning that it had recently been 'put into operation.'[82] He continued his public criticism of the term 'electronic brain' as he had done since his interview for *The Daily Telegraph* and letter to *The Times* in early November 1946. But it was after the publication of Hartree's book in 1949 that Turing began to quote and discuss 'Lady Lovelace's objection' (1950) or 'Lady Lovelace's dictum' (1951a). Hartree drew attention to Lady Lovelace's views in the following way:

> Some of her comments sound remarkably modern. One is very appropriate to a discussion there was in England which arose from a tendency, even in the more responsible press, to use the term "electronic brain" for equipment such as electronic calculating machines, automatic pilots for aircraft, etc. I considered it necessary to protest against this usage [Hartree, D. R. The Times (London), Nov. 7, 1946], as the term would

suggest to the layman that equipment of this kind could 'think for itself,' whereas this is just what it cannot do; all the thinking has to be done beforehand by the designer and by the operator who provides the operating instructions for the particular problem; all the machine can do is to follow these instructions exactly, and this is true even though they involve the faculty of 'judgment.' I found afterward that over a hundred years ago, Lady Lovelace had put the point firmly and concisely (C, p. 44): 'The Analytical Engine has no pretensions whatever to originate anything. It can do whatever we know how to order it to perform.'

(Hartree, 1949, p. 70, her italics)

Hartree further resumed this passage in a way that conceded a window for machine learning research:

This does not imply that it may not be possible to construct electronic equipment which will 'think for itself,' or in which, in biological terms, one could set up a conditioned reflex, which would serve as a basis for 'learning.' Whether this is possible in principle or not is a stimulating and exciting question suggested by some of these recent developments ... But it did not seem that the machines constructed or projected at the time had this property.

(Hartree, 1949, p. 70)

This passage would be quoted and discussed at length by (Turing, 1950, pp. 450, 454, 459). However, Turing expressed his intention to pursue machine learning beyond 'reflexes' and 'the action of the lower centres' of the brain at least as early as circa November 19, 1946, when he wrote a letter to Ross Ashby (Turing, 1946). In fact, Hartree's writing in (1949) was in part a reply to Turing.

In November 1946, Hartree was interviewed alongside Turing about the machine (or 'brain') under construction at the National Physical Laboratory near London, the so-called Automatic Computing Engine (ACE). After World War II, Turing was hired to lead the design of the ACE as an implementation of his 1936 concept of the universal machine (NPL, 1946), while Hartree collaborated with Maurice Wilkes on the EDSAC machine at the University of Cambridge (Rope, 2010; Lavington, 2022). On November 8, *The Daily Telegraph* reported a story based on their interviews, with the headline focusing on Hartree's views.[83] Hartree is reported to have said, 'The implications of the machine are so vast that we cannot conceive how they will affect our civilisation.' While Hartree meant the practical applications of scientific computing, Turing represented a different understanding of the potential impact of computing, as the reporter noted: 'Dr Turing, who conceived the idea of [ACE], said that he foresaw the time, possibly in

30 years, when it would be as easy to ask the machine a question as to ask a man.' The contrast between Hartree's and Turing's views was marked. Hartree is also reported to have said in this 1946 interview, in line with his later citations of Ada Lovelace (1815–1852), that 'the machine would always require a great deal of thought on the part of the operator.' He denied 'any notion that [ACE] could ever be a complete substitute for the human brain.' This was postwar Britain, and Hartree saw a link between the possibility of machine thinking and authoritarian regimes. He added, 'The fashion which has sprung up in the last 20 years to decry human reason is a path which leads straight to Nazism.'

Turing does not seem to have paid attention to Hartree's play of the Nazi card. However, he must have felt compelled to respond to Hartree's interpretation and use of Lovelace's words, for soon after their interviews in early November 1946, during his February 1947 lecture on the ACE to the London Mathematical Society, Turing had already defined his line of response. He accepted the premise of the thesis and questioned its conclusion:

> It has been said that computing machines can only carry out the processes that they are instructed to do. This is certainly true in the sense that if they do something other than what they were instructed, then they have just made some mistake. It is also true that the intention in constructing these machines in the first instance is to treat them as slaves, giving them only jobs which have been thought out in detail, jobs such that the user of the machine fully understands what in principle is going on all the time. Up till the present machines have only been used in this way. But is it necessary that they should always be used in such a manner?
>
> (Turing, 1947, pp. 392–393)

Turing noted that the objection raised by Hartree's use of Lovelace's words was strong and could only be met if machines were made to learn for themselves by experience, without having to be redesigned. He said, 'What we want is a machine that can learn from experience,' and went on to say, 'The possibility of letting the machine alter its own instructions provides the mechanism for this' (1947, p. 393). So when Hartree wrote the above passage in 1949, denying that 'the machines constructed or projected at the time had this property' (of learning to think for themselves), he was probably already responding to Turing's comments of February 1947. He may also have been responding to Norbert Wiener's *Cybernetics* (1948), published in October 1948.[84]

Turing's iconic section, 'Learning machines,' is a high moment of his 1950 paper, making up a quarter of it. This section presents his positive views on machine intelligence. We have just seen that it had a historical grounding in his dialogue with Douglas Hartree, which began in early November 1946. Moreover, a detailed chronology of Turing's concept of machine intelligence shows no reference to any notion of 'learning' by Turing prior to early November 1946 (cf. Chapter 3), when he was interviewed by *The Daily Telegraph* alongside Hartree. He had been talking about machine intelligence in general (without mentioning learning) since at least December 1945.[85] The historical record thus suggests that Hartree's use of Lovelace's words may have helped influence the formation of Turing's concept of machine intelligence in terms of learning from experience.

4.5　Turing's exchanges with Michael Polanyi

Hungarian-born Michael Polanyi (1913–1976) left Nazi Germany for England in 1933 and became an FRS in 1944 (Wigner and Hodgkin, 1977). In 1948, while associated with the Department of Philosophy and with some support from Professor of Philosophy Dorothy Emmet, he was offered a position as the new chair of Social Studies at the University of Manchester. Emmet was an Alfred Whitehead scholar (cf. Swinton, 2019, pp. 87–90). Emmet and Polanyi were interested in the public discussion of science and society in the postwar period and paid attention to the debate over the new computing machines or 'electronic brains.' In this context, they invited Turing, Newman, Jefferson, and others to a seminar on 'The Mind and the Computing Machine,' to be held in the Philosophy Department on October 27, 1949. This would indeed be a pivotal event. We know about it mostly from the minutes that have survived (Turing et al., 1949). I will cover what I see as Polanyi's key interventions that challenged Turing.

　　The seminar had two sessions. The first session was chaired by Polanyi, who read a text entitled 'Can the mind be represented by a machine? Notes for discussion on 27th October 1949,' which he had prepared and given to Newman and Turing several weeks before the meeting.[86] Essentially, Polanyi claimed that humans can solve problems that machines cannot. He tried to support his argument with Gödel's incompleteness theorems. In what survives of the first session of the seminar, we can read the following:

NEWMAN TO POLANYI: The Gödel extra-system instances are produced according to a definite rule, and so can be produced by a machine. The mind/machine problem cannot be solved logically; it must rest on a belief that a machine cannot do anything radically new, to be worked on experimentally. The interesting thing to ask is whether a machine could produce the original Gödel paper, which seems to require an original set of syntheses.

TURING: emphasises the importance of the universal machine, capable of turning itself into any other machine.

POLANYI: emphasises the Semantic Function, as outside the formalis-
able system.

(Turing et al., 1949)

This suggests that Newman, like Turing, believed that 'the mind/machine
problem' could only be solved empirically. Moreover, Newman shifted the
discussion from Polanyi's Gödelian argument to Hartree's thesis associated with Lovelace's words. Both Turing and Newman seem to have tried
to extract some philosophical substance from Polanyi's argument. In particular, Newman related the problem of 'produc[ing] the original Gödel
paper' to what Turing would call in 1950 Lady Lovelace's objection, i.e.,
the question of whether a machine could 'do anything radically new.' In
fact, this connection had been suggested by Turing himself since his lecture on the ACE in February 1947. Using the concept of machine learning, Turing answered both the (then unnamed) objection of Lady Lovelace
(Turing, 1947, pp. 392–393) and the mathematical objection based on
Gödel's argument (pp. 393–394). Polanyi's appeal to a 'semantic function' would continue into the second session of the seminar, chaired by
Dorothy Emmet, and lead to new exchanges with Turing. At some point,
Turing is reported to have presented a distinction to Polanyi, who replied
as follows:

TURING: declares he will try to get back to the point: he was thinking of the kind of machine which takes problems as objectives, and
the rules by which it deals with the problems are different from
the objective. Cf. Polanyi's distinction between mechanically following rules about which you know nothing, and rules about which
you know.

POLANYI: tries to identify rules of the logical system with the rules
which determine our own behaviour, and these are quite different things.

(Turing et al., 1949)

Here lies the motivation for Turing's formulation and response to 'The argument from informality of behaviour' (1950, p. 451). Now, writing nine
years after the October 1949 seminar in Manchester, Polanyi gave this even
more valuable piece of historical information:

A. M. Turing has shown[87] that it is possible to devise a machine which
will both construct and assert as new axioms an indefinite sequence of
Gödelian sentences. Any heuristic process of a routine character — for
which in the deductive sciences the Gödelian process is an example —
could likewise be carried out automatically. A routine game of chess can

be played automatically by a machine, and indeed, all arts can be performed automatically to the extent to which the rules of the art can be
specified.

(Polanyi, 1958, p. 261)

One of the key historical facts that Polanyi provides here is that in October
1949, Turing was still referring to the game of chess as an intellectual task
to illustrate and test for machine intelligence.

Moreover, when we combine what Polanyi is reported to have said in the
notes of the 1949 seminar with what he wrote years later in 1958 (see both
quotations above), we find that Polanyi himself responded to Turing by
classifying chess as an art that 'can be performed automatically' because its
rules 'can be specified.' Thus, in October 1949, Turing saw that his reference
to machine chess was unimpressive for the Manchester philosophers. In
fact, in his report 'Intelligent machinery,' written in the summer of 1948,
Turing had discussed a compromise between the most convenient and the
most impressive intellectual fields:

Of the above possible fields [including 'various games e.g. chess'] the
learning of languages would be the most impressive, since it is the most
human of these activities. This field seems however to depend rather too
much on sense organs and locomotion to be feasible.

(Turing, 1948, p. 421).

At the end of his 1948 report, Turing presented an imitation test for
machine intelligence based on the game of chess. After considering the
strengths and weaknesses of various intellectual tasks and fields for illustrating, developing, and testing machine intelligence, he chose chess. It
had been an important part of his empirical basis for machine intelligence since his report of December 1945 for the NPL (1945, p. 389),
his lecture of February 1947 (1947, p. 393), and his report of summer
1948 report cited above, and survived at least until the Manchester seminar of October 1949, as we can see from Polanyi's report. Then, in his
seminal *Mind* paper, written in late 1949 and early 1950, he replaced
chess, his well-established task, with a viva voce test in the field of language learning. We can now revisit the question: why did he make this
move?

My proposed answer is the following. As we have seen from his
exchanges with Polanyi, he saw that chess would not serve his purpose,
which, according to Robin Gandy, was to persuade 'philosophers and
mathematicians and scientists to take seriously the fact that computers were
not merely calculating engines but were capable of behavior which must be
accounted as intelligent.'

In any case, at the end of his 1950 paper, Turing reconsidered this position:

> We may hope that machines will eventually compete with men in all purely intellectual fields. But which are the best ones to start with? Even this is a difficult decision. Many people think that a very abstract activity, like the playing of chess, would be best. It can also be maintained that it is best to provide the machine with the best sense organs that money can buy and then teach it to understand and speak English. This process could follow the normal teaching of a child. Things would be pointed out and named. Again I do not know what the right answer is, but I think both approaches should be tried.
>
> (Turing, 1950, p. 460).

Turing was not sure. As he had already suggested in the summer of 1948, he was hesitant because of the cost of providing 'child machines' with 'sense organs and locomotion' so that they could learn a language. Clearly, chess was more convenient for use in initial experiments in the early 1950s, while conversational question-answering was still an imaginary experiment, though preferable for persuasion about human intelligence. In further discussions with essentially the same interlocutors, Turing reiterated his proposal of various forms of viva voce examination to test for machine intelligence in 1951 (Turing, 1951a, p. 484) and 1952 (Turing et al., 1952, p. 495), although he again reconsidered the virtues of chess in (Turing, 1953, p. 569), which was written in mid-1951 (cf. the previous chapter, §3.2.7).

It turns out that Turing's chronological development of machine intelligence shows that he never claimed a single and special form of ('species') testing as a crucial experiment for human-level machine intelligence. Instead, he acknowledged the existence of several 'imitation tests.'[88] If Turing eventually emphasized conversation over chess to illustrate, develop, and test machine intelligence, he did so, if for no other reason, under the influence of Polanyi's criticism that chess is an art that 'can be performed automatically' because its rules 'can be specified.'

So far we have seen that in his 1950 paper, Turing responded to the criticisms of Hartree (November 1946 and June 1949) and Polanyi (October 1949). In particular, we have seen that Turing left chess as a kind of second option to embody an intelligence test after Polanyi's criticism. It turns out that his positive adoption of linguistic performance instead of chess playing also has historical roots in his exchanges with Jefferson in the same period (June to December 1949). As we will now see, Jefferson became Turing's main intellectual opponent.

4.6   Turing's exchanges with Geoffrey Jefferson

Geoffrey Jefferson (1886–1961), then Professor of Neurosurgery at the University of Manchester and an FRS since 1947 (cf. Walshe, 1961), read his Lister Oration in London on June 9, 1949, which was published two weeks later in the *British Medical Journal* (1949a). Jefferson laid down criteria and demands without which he could not 'agree that machine equals brain' (p. 1110). He titled his lecture 'The mind of mechanical man' in response to Norbert Wiener's *Cybernetics* (1948) and to the various digital computing projects in the UK and the US, especially the one Turing was involved in at their University of Manchester. A reporter from *The Times* covered Jefferson's memorial and highlighted one of Jefferson's strongest observations, which was quoted the next day (June 10, 1949) under the headline 'No mind for mechanical man:'

> [N]ot until a machine can write a sonnet or a concerto because of thoughts and emotions felt, and not by the chance fall of symbols, could we agree that machine equals brain — that is, not only write it but know that it had written it. No mechanism could feel (and not merely artificially signal, an easy contrivance) pleasure at its successes, grief when its valves fuse, be warmed by flattery, be made miserable by its mistakes, be charmed by sex, be angry or miserable when it cannot get what it wants.[89]
>
> (cf. also Jefferson, 1949a, p. 1110)

The reporter asked the Computing Laboratory at the University of Manchester for a reply to Jefferson's claims. Once asked,[90] Turing made a sharp and witty rebuttal. On the next day (June 11, 1949), he was quoted in the newspaper under headline 'Calculus to Sonnet:'

> Mr. Turing said yesterday: 'This is only a foretaste of what is to come, and only the shadow of what is going to be. We have to have some experience with the machine before we really know its capabilities. It may take years before we settle down to the new possibilities, but I do not see why it should not enter any one of the fields normally covered by the human intellect, and eventually compete on equal terms.' 'I do not think you can even draw the line about sonnets, though the comparison is perhaps a little bit unfair because a sonnet written by a machine will be better appreciated by another machine.'
>
> Mr. Turing added that the University was really interested in the investigation of the possibilities of machines for their own sake. Their research would be directed to finding the degree of intellectual activity of which a machine was capable, and to what extent it could think for itself. News

of the experiments was disclosed by Professor Jefferson in the Lister Oration reported in The Times yesterday.[91]

When Jefferson's Lister Oration appeared in the BMJ (June 25), Turing was addressed in a warning note from the editorial that opened the edition:

> Mr. A. W. Turing [*sic*], who is one of the mathematicians in charge of the Manchester 'mechanical brain,' said in an interview with The Times (June 11) that he did not exclude the possibility that a machine might produce a sonnet, though it might require another machine to appreciate it. Probably he did not mean this to be taken too seriously ...
>
> (BMJ, 1949, p. 1129)

Turing would push back in 1950. It turns out that an imaginary sonnet-writing machine is exactly what he presented in his 1950 paper. This is evidence that not only Polanyi's negative point about chess but also Jefferson's positive demand for sonnets influenced Turing's shift from chess to conversation as the intellectual task to test machine intelligence. Turing quoted Jefferson's demands and addressed Jefferson directly:

> I am sure that Professor Jefferson does not wish to adopt the extreme and solipsist point of view. Probably he would be quite willing to accept the imitation game as a test. The game (with the player B omitted) is frequently used in practice under the name of *viva voce* to discover whether some one really understands something or has 'learnt it parrot fashion.'
>
> (Turing, 1950, p. 446, his emphasis)

In fact, Jefferson's Lister Oration was a bold critique of the Turing-Wiener analogy between the new electronic computing machines and the human brain. He argued against the idea that machines could think, even linking it to 'political' and 'religious' issues. He urged that 'the concept of thinking like machines lends itself to certain political dogmas inimical to man's happiness' and 'erodes religious beliefs that have been mainstays of social conduct' (1949a, p. 1107). The influence of Jefferson's text on Turing's 1950 paper is material and substantial. While this general point should not surprise Turing scholars, Jefferson's influence on Turing's 1950 paper has not yet been fully appreciated.

Very recent evidence suggests that another edition of the seminar took place in December 1949. Jonathan Swinton found a Christmas 1949 postcard sent to cybernetician Warren McCulloch (1898–1969), then in Chicago, by a Jules Bogue, a chemical industrialist and a neighbor of Max Newman at the time, who found his way into the meeting:

> I wish you [McCulloch] had been with us a few days ago we had an amusing evening discussion with Thuring [*sic*], Wiliams, Max Newman, Polyani [*sic*], Jefferson, j z Young & myself. An electronic analyser and a digital computer (universal type) might have sorted the arguments out a bit.[92]

Some chaos was noted in the arguments during the discussion in December 1949, which may explain Turing's aim to propose the imitation game 'as a basis for discussion' (1950, p. 445).

Now, I have observed that this finding of Swinton's correlates with what Jefferson related in a letter after Turing's death. Jefferson described an event when Turing would have come to his house to talk to Professor J.Z. Young (1907–1997) and himself over dinner after a meeting in the Philosophy Department. The key information that Jefferson gave was that after midnight, Turing went to ride home on his bicycle 'through the same winter's rain' (Irvine, 1959, p. xx). So, if we take Jefferson's word at face value, that meeting cannot have been the seminar of October 27, 1949 (in the fall) and must have taken place in late December (in the winter) close to Christmas Eve. Also, given that the minutes of the October 1949 edition (Turing et al., 1949) do not show any exchange between Jefferson and Turing, it must have been at this December meeting (extending late into the night at Jefferson's house) that they had their most lively exchanges. This must have been when Jefferson drew Turing's attention to his Lister Oration. In fact, note the topic of the viva-voce examination that Turing presents to illustrate the imitation game:

> Interrogator: In the first line of your sonnet which reads 'Shall I compare thee to a summer's day', would not 'a spring day' do as well or better?
> Witness: It wouldn't scan.
> Interrogator: How about 'a winter's day'. That would scan all right.
> Witness: Yes, but nobody wants to be compared to a winter's day.
> Interrogator: Would you say Mr. Pickwick reminded you of Christmas?
> Witness: In a way.
> Interrogator: Yet Christmas is a winter's day, and I do not think Mr. Pickwick would mind the comparison.
> Witness: I don't think you're serious. By a winter's day, one means a typical winter's day, rather than a special one like Christmas.

And so on. What would Professor Jefferson say if the sonnet-writing machine was able to answer like this in the *viva voce*?

(Turing, 1950, p. 446)

Turing's imaginary sonnet-writing machine is asked about the first line of its sonnet, 'Shall I compare thee to a summer's day?' as in Shakespeare's sonnet, and both Shakespeare and the demand for a sonnet-writing machine appear in the climax of Jefferson's Lister Oration. Further, the topic of the machine's interrogation shifts to a winter's day in general and Christmas Eve in particular. The connection with the historical event of Turing's dinner at Jefferson's house close to the Christmas Eve of 1949 is striking.

It was probably in that December 1949 night that Jefferson gave to Turing an offprint of his Lister Oration (1949a) and his paper on Descartes (1949b), which remained in Turing's possession and was given to the King's College Archives, Cambridge, after Turing's death. The Archive's catalog entry describes it as having 'annotations by AMT (Alan Turing).'[93] Darren Abramson (2011) pointed this out (p. 548), reporting two heavy markings on the offprint of Jefferson's Lister Oration, which I confirmed with my own eyes. This provides material evidence that Turing read Jefferson's text and was impressed by certain parts of it. Turing marked two passages with a pink pencil: Jefferson's demands that appeared in *The Times*, as we have seen above, and Jefferson's exposition of René Descartes' 1637 *Discourse on Method*, Part V (1949a, p. 1106). The latter presented Descartes' sensible image of a viva-voce examination to distinguish humans from machines and other animals, no matter how ingenious their imitation of human behavior might appear to be at first glance. Toward the end of his text, Jefferson returned to Descartes to propose speech as the most distinctive intellectual faculty of 'man' as opposed to 'the highest animal' (p. 1109), and further demanded that thinking machines should be able to write a sonnet 'because of thoughts and emotions felt' (p. 1110). Thus, by imagining in his imitation game a machine being questioned about a sonnet it had composed, Turing addressed both of Jefferson's demands — writing a sonnet in Shakespeare's standard and passing a Descartes-like viva-voce test about it — at once.

Now, Jefferson made another move that to my knowledge has never been observed in the secondary literature and yet is crucial to understanding Turing's test. Jefferson offered Turing a second image, and this one was no less striking than the other image. Jefferson referred to 'sex hormones' as a distinguishing feature of the behavior of 'animals' and 'men,' as opposed to 'modern automata' (1949a, p. 1107). As part of this connection, he referred to Grey Walter's iconic electromechanical tortoises:

[It] should be possible to construct a simple animal such as a tortoise (as Grey Walter ingeniously proposed) that would show by its movements that it disliked bright lights, cold, and damp, and be apparently frightened by loud noises, moving towards or away from such stimuli as its receptors were capable of responding to. In a favourable situation the behaviour of such a toy could appear to be very lifelike — so much so that a good demonstrator might cause the credulous to exclaim 'This is indeed a tortoise.' I imagine, however, that another tortoise would quickly find it a puzzling companion and a disappointing mate.

(Jefferson, 1949a, p. 1107)

Jefferson further remarked that 'neither animals nor men can be explained by studying nervous mechanics in isolation, so complicated are they by endocrines, so coloured is thought by emotion.' He then stated that '[s]ex hormones introduce peculiarities of behaviour often as inexplicable as they are impressive' (*ibid.*). In short, Jefferson suggested that machines could not exhibit enough peculiarities of behavior to be able to imitate the actions of animals or 'men' because they have no sex hormones. A machine would give itself away and be found to be 'a puzzling companion and a disappointing mate.' Jefferson thus suggested that the physiology of sex hormones is causally related to interesting behavior, meaning gendered behavior.

In Jefferson's passage quoted by *The Times* in June 1949, he notably argues that machines should be able to have emotional reactions in general and be capable of being 'charmed by sex' in particular if they can be said to think. In his 1950 paper, Turing addressed this in his discussion of objection (5) 'Arguments from various disabilities' (p. 447). Among other nonobvious things, he considered the ability to 'fall in love' and 'make someone fall in love with it' as within the reach of machines. Although Turing did not address Jefferson's tortoise challenge directly, one may note that for a machine to not be a puzzling companion and a disappointing mate in the sense of Jefferson, it must be able to learn and imitate gender. Turing addressed the tortoise challenge more subtly, one might say, in the very design of his imitation game. He modified the viva-voce examination proposed by Descartes (1637) in a few key aspects. In Descartes' test, there were only two participants: the one contestant entity — an animal or machine — and the human interrogator questioning it. Turing, having introduced an arrangement for blind communication to control for bias, introduced another arrangement for a third player, player B, who is supposed 'to help the interrogator' in making the right decision. Player B is meant to be gendered and to sit side by side with player A, that is, to serve as a baseline model of its gender performance in the unrestricted conversation conducted by the examiner. Now, let us recall Jefferson's argument about the influence of sex hormones in the production of peculiarities of behavior in 'animals' and 'men.' Jefferson presented the image of an electromechanical tortoise that is put side by side with an actual tortoise. Suppose by analogy that we consider, as Jefferson suggested in the title of his Lister Oration, a 'mechanical man' side by side with an actual woman or man. Turing's question is thus posed: without being able to see, touch or hear the two, would one be able to tell them apart? Or would the machine, as predicted by Jefferson in his critique of Walter's electromechanical tortoises, be quickly found to be 'a puzzling companion and a disappointing mate'? If the thesis that sex hormones are crucial to produce interesting behavior was also at stake, then Descartes's language test by itself, even if fixed by Turing's arrangement for blind communication, would fall short at satisfying one of Jefferson's conditions for machine intelligence. It would have to be extended along the lines of Turing's imitation game. If player A can imitate the required gender sufficiently well, then it will showcase not only human intelligence in general but also the 'peculiarities of behaviour' that according to Jefferson would be rendered by specific (male/female) 'sex hormones.'

Now, as a homosexual man, Turing must have been sure that gendered behavior was not causally determined by male/female sex hormones and could be learned and imitated. Therefore, in his subtle rebuttal to Jefferson, Turing seems to be arguing that machines could learn and imitate gender, yes, they could. It is astonishing to observe that Turing designed his imitation game in late 1949 and early 1950 to challenge Jefferson's thesis, and two years later he would be pushed to inject female hormones to control his libido.

4.7   Conclusion

This chapter has drawn attention to the mind-machine controversy in England (1946-1950), which led Turing to propose his famous imitation game or test for machine intelligence. I have provided original answers to key exegetical and historical questions that have not been adequately addressed.

Turing spent several years — possibly from *c.* 1941, and at least from 1945 until the end of 1949 — working with chess as a task to illustrate, develop, and test for machine intelligence. At least since his indirect dialogue with Hartree about the cognitive capabilities and limitations of the ACE in late 1946, Turing had been thinking about building a machine that could learn to play chess from experience. His goal was to establish a concept of machine intelligence that would not fall prey to the Hartree's interpretation of Lovelace's dictum. Learning from experience was Turing's answer to Hartree's criticism and that could be illustrated quite well in a game of chess. However, in October 1949, his argument based on chess received criticism from Polanyi, who was unimpressed and argued that chess was an art that 'can be performed automatically,' because its rules 'can be specified.' No less important, at the end of the year, Jefferson drew Turing's attention to his Lister Oration. In his text, Jefferson shed light on Descartes' proposal of a viva-voce examination to distinguish humans from machines and other animals and pointed to speech as highest form of human intelligence. Jefferson's emphasis on language was such that the climax of his Lister Oration was to demand that thinking machines be able to write a sonnet. He further required that the machine's linguistic performance be tied to emotions. Additionally, Jefferson referred to Grey Walter's iconic electromechanical tortoises and suggested that machines could not exhibit enough peculiarities of behavior to be able to imitate the actions of animals or 'men' because they lack sex hormones. A machine would give itself away and be found to be 'a puzzling companion and a disappointing mate.' Jefferson was suggesting that the physiology of sex hormones is causally related to gendered behavior. Turing challenged Jefferson's position with an irreverent adaptation of Descartes' test. I suggest that Turing's dialogue with Jefferson's Lister Oration provides evidence that Turing did propose gender learning and imitation as one of his various imitation tests for machine intelligence[94] and that this was probably done in response to Jefferson.

In summary, I have explained Robin Gandy's anecdote about the purpose of the Turing test and highlighted Turing's most notable interlocutors — the 'philosophers, mathematicians, and scientists' whom Turing 'sought to persuade' about the cognitive capabilities of digital computers. I have argued that Turing's direct and indirect discussion with these three thinkers, Hartree, Polanyi, and Jefferson, is key to any exegesis of Turing's 1950 paper and to an understanding of the conceptual problems he sought to solve with his proposal of the imitation game, which became widely known as his test for machine intelligence.

5   The Turing Test Is a Thought Experiment

Turing's imitation game and test has been studied and run as a controlled experiment and found to be underspecified and poorly designed. It has also been defended and continues to attract interest as a test of artificial intelligence. Scientists and philosophers alike acknowledge that the current status of the test is at odds with the intellectual standards of Turing's works. This chapter refers to this as the Turing Test Dilemma, after noting that the test has long been debated and is still widely regarded as either too bad or too good to be a valuable practical experiment. An argument is presented that resolves the dilemma by reconstructing the Turing test as a thought experiment in the modern scientific tradition. It is argued that Turing's exposition of his test satisfies Mach's characterization of 'the basic method of thought experiments' and that Turing made critical and heuristic uses of his test. He methodically varies its design to address specific challenges posed to him by other thinkers. He also illustrates a property of the phenomenon of intelligence and suggests a hypothesis about machine learning. This reconstruction of the Turing test provides a rapprochement to the conflicting views of its value in the literature.[95]

5.1   The Turing Test Dilemma

Alan Turing opened his seminal paper by proposing to replace the question 'can machines think?,' which he deemed 'too meaningless to deserve discussion' (1950, p. 442). The new question, considered to have a 'more accurate form,' would be based on what Turing called the 'imitation game,' and later in the same text, his 'test.'[96] Essentially, according to different interpretations of the various versions of the test, the machine must be able to imitate stereotypes of a woman, a man, or a human, beside a true representative of the kind, to deceive a human interrogator about its true nature. The new question is whether the interrogator, at a distance and having no physical contact whatsoever, would be able to distinguish the machine from the genuine individual through a conversation game. If not, the machine must be considered intelligent.

However, details about this new question and the exact settings for its evaluation (duration, number of test runs, scoring protocol, characterization of the players and interrogators) slip through Turing's 1950 text in

DOI: 10.4324/9781003300267-5

a sequence of variations that defies interpretation. Two different versions have been identified (Sterrett, 2000; Traiger, 2000) and have been referred to as the 'Original Imitation Game' (read from pp. 433–434 in Turing, 1950) and the 'Standard Turing Test' (read from p. 442). There is significant disagreement on how the two passages should be read. Some authors acknowledged the presence of a 'gender test' in the first passage (Genova, 1994; Hayes and Ford, 1995). Others considered it to serve as a scoring protocol for a nongendered test read from the second passage (Copeland, 2004, p. 436; Proudfoot, 2013, p. 395). Others have disregarded any form of gender imitation and read from the second passage instead, a standalone Standard Turing Test (Moor, 1976; Dennett, 1984; Piccinini, 2000; Moor, 2001; Shieber, 2007), which turns out to be the most popularized version of the test. According to it, the game tests the machine's capability of giving sensible answers to questions, both complex and simple, indistinguishably from a human in an unrestricted conversation game conducted by the interrogator.

This chapter refers to the 'imitation game,' 'Turing's test(s),' and the 'Turing test' without committing to a specific passage from Turing's texts; instead, it considers a conflation of several passages that will be examined in due course (Section 5.3). Beyond existing disputes about which Turing test is best for artificial intelligence (AI), this chapter will characterize influential positions on the 'Turing Test Dilemma,' which asks whether the Turing test is a valuable experiment for AI. Preliminaries are presented (Section 5.1.1), and two main positions are identified (Sections 5.1.2 and 5.1.3) relative to the two horns of this dilemma.

5.1.1   Preliminaries: the practical Turing tests

First, it is worth noting that there have been several attempts at running the test as a controlled experiment. Two such ventures received much attention — the 1991 and the 2014 editions of the Loebner Prize Competition held in Boston (Epstein, 1992) and at the Royal Society in London (Warwick and Shah, 2015). The top-ranked program in 1991 was the PC Therapist, developed by a psychology graduate turned computer programmer. It was inspired by ELIZA, a trick developed in the mid-1960s by Joseph Weizenbaum (1966) to imitate a 'person-centered' (Rogerian) therapist by regurgitating the patient's own words and phrases in a simulation of understanding. The top-ranked program in 2014 was Eugene Goostman, the simulation of a Ukrainian boy that claimed to be restricted by his acquired culture at thirteen and his use of English as a second language.

Based on his first-person experience with the former, Stuart Shieber (1994a) provided a rigorous and comprehensive analysis of the problem of implementing a practical Turing test. To have its coherence preserved,

Shieber remarked, the Turing test could not be restricted in its domain (it must be open for any conversation topic) or task (it must be open for any question). Shieber recommended that practical Turing tests should not be run until the standard of AI gets close to the high standards required by the test.[97] However, the Loebner Prize has been continued all the same. The 2014 edition was organized by Kevin Warwick and Huma Shah, who had been experimenting with practical implementations of the Turing test for several years (2016). Warwick and Shah (2015) announced the Eugene Goostman program as being 'the first to pass the Turing test.' They argued that their 2014 implementation of the Turing test was unrestricted 'as set out by Alan Turing.' Having received criticism from Moshe Vardi (2014), Shah and Warwick (2015) presented evidence that the acclaimed program does seem indistinguishable from humans in conversation. Yet, Vardi's rejoinder ran: 'The details of this 2014 Turing Test experiment only reinforces my judgment that the Turing Test says little about machine intelligence' (*ibid.*).

5.1.2   *The negative answer: the Turing test is too flawed to be a valuable experiment for AI*

Given the relatively good performances of obviously unintelligent machines in practical Turing tests, the scientific community seems to have mostly opted to dismiss the test, which would have been revealed to be 'just a game' (Vardi, 2014) or 'highly gameable' (Marcus et al., 2016). This had been discussed by an earlier influential address given by Hayes and Ford (1995), who declared to have tried to 'take Turing seriously.' They acknowledged that Turing's test 'has been with AI since its inception, and has always partly defined the field.' Further, they recollected, '[s]ome AI pioneers seriously adopted it as a long-range goal, and some long-standing research programs are still guided by it.' They suggested that scientists abandon the goal of constructing a 'mechanical transvestite.' They also referred to the practical Turing tests, which would have shown that the test has plenty of ambiguities, flaws, and gaps in its design. Further, it would be a biased and even circular test, the standard of which would be elusive, and it would be unable to detect anything. Accordingly, the Turing test should be rejected and moved 'from the textbooks to the history books' (Hayes and Ford, 1995). Bringsjord et al. (2001) emphasized that attempts to build computational systems able to pass restricted versions of the test have devolved into shallow symbol manipulation designed to fool people and concluded that 'the problem is fundamental: the structure of the [test] is such as to cultivate tricksters.' In summary, '[c]onsidering the importance Turing's Imitation Game has assumed,' Drew McDermott wrote in (2014), 'it is a pity he was not clearer about what the game was exactly.'

In general, critics of the Turing test answer 'no' to the Turing Test Dilemma. According to them, it is unfortunate that the test is an underspecified and poorly designed experiment. However, that position must face the first horn of the dilemma: it is at odds with the intellectual standards of Turing's works (Newman, 1955). Further, if the test is so bad, why has it been defended and attracted so much interest? Would that be due to Turing's credentials alone?

5.1.3   *The positive answer: the Turing test is too good to be abandoned as an experiment for AI*

The Turing test has been defended before and since its early 1990s practical implementations, primarily by AI philosophers. James Moor (1976) was the first to emphasize the generality of the test and to advocate its use in unrestricted experiments (pp. 249–250). Daniel Dennett (1984) noted that the test comes from a long philosophical tradition ('[p]erhaps he was inspired by Descartes,' p. 297) and observed that it is general enough to subsume several specific intellectual tasks at once. Dennett argued that 'the Turing test, conceived as he conceived it, is (as he thought) plenty strong enough as a test of thinking,' and provoked: 'I defy anyone to improve upon it' (p. 297). He argued that it is a convenient sufficient condition (a 'quick probe,' p. 298) for confirming the presence of a human-level AI. After the first practical Turing tests, Dennett (1997) regretted that the Turing test 'requires too much Disney and not enough science' and that it 'is too difficult for the real world' (p. 315). Jack Copeland (2000b) rejoined: 'It is often claimed that Turing was insufficiently specific in his description of his test' (p. 530). 'A machine emulates the brain,' Copeland clarified, 'if it plays the imitation game successfully come what may, with no field of human endeavour barred, and for any length of time commensurate with the human lifespan.' Concerning the difficulties of implementing such an unrestricted experiment, he suggested that the solution lies in sampling: 'Any test short enough to be practicable is but a sampling of this ongoing situation.' Shieber (2007) presented a statistical-proof scheme to substantiate the inferential status of the Turing test as a sufficient condition for intelligence, and arguably it could be adapted along the lines suggested by Copeland. However, despite the availability of such an elegant mathematical device, according to Turing, the test must rely on the judgment of 'an average interrogator' (1950, p. 442) or of 'a jury, who should not be expert about machines' (1952, p. 495), and such judgments can be flawed.[98]

In general, supporters of the Turing test answer 'yes' to the Turing Test Dilemma. They hold that, in its original (unrestricted) form, the test is not comparable with the restricted practical Turing tests run so far and is too good to be abandoned as an AI experiment. However, this leads to the second horn of the dilemma: if the test cannot be supplanted, will the success of AI science depend on the chances of average human interrogators against increasingly elaborate, yet still unintelligent, chatbots in unrestricted tests? In any case, does running repeated unrestricted Turing tests bring value to AI?

Altogether, taking the dilemma by any one of the two horns, no simple and general explanation of the Turing test seems available to deal with the other horn.

5.2   Argument sketch

This chapter argues that the Turing Test Dilemma can be solved by reconstructing the test as a thought experiment in the modern scientific tradition. No study of the Turing test appears to have ever reconstructed it as a thought experiment.

A core criticism of the test's value as an AI experiment is that Turing would not have specified exact settings for implementing it, whose design would turn out to be poor and imprecise. This view is evidenced, for instance, by the existence of two widely acknowledged and yet heterogeneous readings of the test: the Original Imitation Game and the Standard Turing Test. However, this chapter will argue that Turing's presentation of his test (Section 5.3) satisfies what Ernst Mach called 'the basic method of thought experiments' (Section 5.4), characterized by a continuous variation of experimental conditions (1897).[99] 'By astute handling of this procedure,' Mach observed, 'we may reach cases that at first blush seem rather different, that is to generalisation of the point of view.' Showing that Turing's presentation of his test satisfies Mach's observations establishes that the Turing test can be understood as a thought experiment in the modern scientific tradition, had Turing been aware of it or not.[100] Accordingly, the critique that the test is an underspecified and flawed experiment can be rebutted by showing the rich methodological structure in Turing's exposition of his imitation game and test.

Also in support of understanding Turing's proposal within the scientific tradition, this chapter will reconstruct the Turing test as a thought experiment serving both critical and heuristic uses (Sections 5.5 and 5.6). Karl Popper (1959) presented a discussion of 'apologetical,' 'critical' and 'heuristic' uses of 'imaginary experiments' (pp. 465–466). Popper found in Galileo's criticism of Aristotle's theory of motion in the context of his polemic with peripatetic philosophers the paradigmatic case of the *critical* use of thought experiments. Similarly, Turing's critical use of his test addressed and posed severe problems to opposing theories of intelligence presented to Turing by his intellectual opponents in the context of controversy. In particular, seeking conceptual change on the meaning of the words 'machine' and 'think,' Turing tried to expose a paradox in a theory of intelligence that tied logical kind to physical kind, which had been presented to him by a contender as will be shown later. This satisfies Thomas Kuhn's conception of the function of thought experiments (1964). Popper also pointed out Einstein's experiment of the accelerated lift as a paradigmatic case of the *heuristic* use of thought experiments as 'it illustrates the local equivalence of acceleration and gravity, and it suggests that light rays in a gravitational field may proceed on curved paths.' According to Popper, therefore, the heuristic use illustrates a property of the studied phenomenon and suggests a related hypothesis. The reconstruction of Turing's heuristic use of his test will conform to that scheme. The Turing test illustrates that the perception of intelligence is emotional, and it suggests the hypothesis that a learning machine may be created simple and educated naturally, without reboots or special coaching, to play the imitation game well.

The reconstruction of Turing's critical and heuristic uses of his test will emphasize how it increases understanding of the question 'can machines think?' and prepares for related practical experiments. Attention will be drawn to how the imitation game accomplishes its epistemic goals through its design and not by its execution.[101] Overall, the reconstruction of the test as a thought experiment will provide a rapprochement to the conflicting views on the value of the Turing test for AI and can ultimately end the Turing Test Dilemma as a two-horned issue.

This argument sketch summarizes this chapter's contributions to advancing a crucial debate on the conceptual foundations of AI and machine learning. The remainder presents the complete argument in detail. The key points will be revisited at the end (Section 5.7).

5.3　Turing's presentation of his test

Turing's presentation of his test is studied by emphasizing how he varies the conditions of his test (Section 5.3.1). Then, the methodological structure of Turing's exposition is outlined (Section 5.3.2).

5.3.1　*Turing's variation of the conditions of his test*

To replace the question (Q) 'can machines think?,' Turing introduced his imitation game:

> The new form of the problem can be described in terms of a game which we call the 'imitation game'. It is played with three people, a man (A), a woman (B), and an interrogator (C) who may be of either sex. The interrogator stays in a room apart from the other two. The object of the game for the interrogator is to determine which of the other two is the

man and which is the woman. He knows them by labels X and Y, and at the end of the game he says either 'X is A and Y is B' or 'X is B and Y is A' ... It is A's object in the game to try and cause C to make the wrong identification ... The object of the game for the third player (B) is to help the interrogator.

We now ask the question, 'What will happen when a machine takes the part of A in this game?' Will the interrogator decide wrongly as often when the game is played like this as he does when the game is played between a man and a woman? These questions replace our original, 'Can machines think?'

(Turing, 1950, pp. 433–434)

The substitute question (Q'), therefore, was based on Turing's imitation game and test. It has been referred to, as mentioned, as the Original Imitation Game. Turing then illustrated a few queries that the interrogator could make and suggested that all communication between the interrogator and the participants should be teletyped to neutralize signals such as tone of voice. Structurally, Turing's first presentation of the game relates two variants: a baseline *man-imitates-woman* game and a *machine-imitates-woman* game. Results of the latter are supposed to be compared with the results of the former. In commenting on practical Turing tests, Copeland (2004) argued that this comparison of results is a scoring protocol (p. 436). However, this misses the point that the comparison performs a conceptual function. It reminds the reader of a common-sense truism—namely, that a man can imitate stereotypes associated with women despite their biological difference.

Turing proceeded to discuss strengths and weaknesses of the new problem and which machines would be concerned in the game. Having introduced digital computers as the kind of machine allowed to take part in the game, he paused and revisited the new problem:

There are already a number of digital computers in working order, and it may be asked, 'Why not try the experiment straight away? It would be easy to satisfy the conditions of the game. A number of interrogators could be used, and statistics compiled to show how often the right identification was given.' The short answer is that we are not asking whether all digital computers would do well in the game nor whether the computers at present available would do well, but whether there are imaginable computers which would do well.

(Turing, 1950, p. 436)

This new formulation can be identified as Q'': are there 'imaginable computers' that could perform well in the imitation game? This reference to an imaginary experiment should not pass by unnoticed. Turing promised to present that question 'in a different light later,' and proceeded to explain a key scientific property of the new digital computers: their universality. He had given a conceptual description of the digital computer as a discrete-state machine. He then used the imitation game to illustrate his point once again:

> Given the table corresponding to [any] discrete state machine it is possible to predict what it will do ... the digital computer could mimic [its] behaviour. The imitation game could then be played with the machine in question (as B) and the mimicking digital computer (as A) and the interrogator would be unable to distinguish them.
>
> (Turing, 1950, p. 441)

Turing further remarked that '[t]his special property of digital computers, that they can mimic any discrete-state machine, is described by saying that they are *universal* machines' (p. 441, no emphasis added). Turing thus used this *machine-imitates-machine* variant of the game to suggest that physical kinds could, in principle, have their logical behavior imitated, as long as the imitating agent was properly qualified for universal computation.

In yet another variation, Turing considered 'again the point raised at the end of §3' (Q''), which he had promised. Now, having explained the science and technology of digital computers and their universality property, he posited:

> It was suggested tentatively that the question, 'Can machines think?' should be replaced by [question Q'', which is also] equivalent to this, 'Let us fix our attention on one particular digital computer C. Is it true that by modifying this computer to have an adequate storage, suitably increasing its speed of action, and providing it with an appropriate programme, C can be made to play satisfactorily the part of A in the imitation game, the part of B being taken by a man?'
>
> (Turing, 1950, p. 442)

This version of the test (Q''') reinstates the A/B/C-player structure in a *machine-imitates-man* game. Turing's reference to 'man' has been generally read as masculine generics. This is the case of the Standard Turing Test, which reads in Turing's passage a *machine-imitates-human* game and discards the baseline man-imitates-woman game as an implicit scoring protocol in question Q'''. The present reconstruction of the Turing test as a thought experiment can end the exegetical problem of whether Turing meant an ungendered human, as will be shown later (Section 5.5.3). In any case, Turing's literal use will be followed for simplicity, and this version will be referred to as a 'machine-imitates-man' game.

Once having considered 'the ground to have been cleared' (p. 442), Turing revisited 'the original form of the problem' (Q) and 'the more accurate form of the question' (Q'''):

> I believe that in about fifty years' time it will be possible to programme computers, with a storage capacity of about 10^9, to make them play the imitation game so well that an average interrogator will not have more than 70 per cent, chance of making the right identification after five minutes of questioning.
>
> (Turing, 1950, p. 442)

Turing guesses an answer to yet another question (Q''''): can a machine of gigabit-storage capacity be programmed to deceive an average interrogator in 30% of the times that it plays the imitation game for five minutes?

In his text (1950), Turing presented research steps that 'should be taken now if the experiment [question Q''''] is to be successful' (p. 455). Therefore, contrary to the view of some commentators that Q'''' is a prediction, and thus it could not rule the test, Turing did suggest that it is a valid version of 'the experiment.' Warwick and Shah (2015) sought to implement the conditions of Q'''' very closely and claimed that the 'Eugene Goostman' chatbot satisfied it. So, thinkers that answer positively to the Turing Test Dilemma either diverge from Turing's original proposal or should not reprobate the claim.

At the end of his text (1950), Turing was unsure about which intellectual field was best to address in a test for machine intelligence. He referred to machines eventually competing with men 'in all purely intellectual fields' and asked (p. 460): 'But which are the best ones to start with?' He pondered that even this 'is a difficult decision' and added: 'Many people think that a very abstract activity, like the playing of chess, would be best. It can also be maintained that it is best to provide the machine with the best sense organs that money can buy, and then teach it to understand and speak English.' In (1948), Turing had discussed kinds of intelligence task to be explored in machine intelligence research (pp. 420–421) and even described an imitation test based on the game of chess, referring to it as 'a rather idealized form of an experiment I have actually done' (p. 431). In (1951a; 1952), he presented yet other versions of his test, having even acknowledged the existence of several 'imitation tests' (cf. note 96). Altogether, Turing presented various imitation tests not only *throughout* his 1950 text, but also *before* and *after* it.

5.3.2   The case-control methodological structure of Turing's various conditions and questions

To replace the original question, Q, Turing posed in his 1950 paper various empirical questions, Q' to Q'''', based on different game variants through varying players A and B while keeping C fixed: (i) man-woman, (ii) machine-woman, (iii) machine-machine and (iv) machine-man and (v) machine as A in the absence of B. Those questions can be generalized as follows:

Question Q^*: could player A imitate intellectual stereotypes associated with player B's type successfully (well enough to deceive player C), despite the physical differences between A's and B's types?

Turing's varied conditions establish two levels of case-control structure. At the intra-game level, A plays the case, and B plays the control. At the inter-game level, the case-control structure alternates as follows. Question Q^* is open concerning the *machine-woman* and the *machine-man* versions of the game, both of which set the case; however, the same question is settled concerning the *man-woman* and the *machine-machine* variants of the game, which set the control. Beyond Turing's rhetorical use of the man-woman variant of the game, it is well known that a man (A) can possibly imitate gender stereotypes associated with a woman (B) successfully, despite their physical difference. Further, regarding the machine-machine variant, it is also known that a digital computer (A), because of its universality property proven by Turing (1936), can successfully imitate any discrete-state machine (B), despite their physical difference.

We may now proceed to analyze Turing's presentation of his test against the backdrop of classical conceptions of thought experiments in the philosophy of science literature.

5.4   Turing's use of the basic method of thought experiments

Turing's presentation of his test satisfies Mach's conception of the basic method of thought experiments, which is variation, continuously if possible. Ernst Mach's characterization of the method is presented (Section 5.4.1) and then compared with Turing's use of it in his exposition of his imitation tests (Section 5.4.2).

5.4.1   Mach's characterization of the method

Mach developed sharp observations and insights on thought experiments throughout his text, which he grounded in countless examples from the history of modern physics, mathematics, and commonsense experience. On the method, he wrote:

[T]he basic method of thought experiments, as with physical experiments, is that of variation. By varying the conditions (continuously if possible), the scope of ideas (expectations) tied to them is extended: by modifying and specializing the conditions we modify and specialize the ideas, making them more determinate, and the two processes alternate.[102]

(Mach, 1897, p. 139)

It is important to note the mutually reinforcing connection he suggested between extending 'the scope of ideas (expectations)' and 'varying the conditions,'[103] where variation means 'modifying and specializing,' continuously 'if possible.' Mach illustrated his point through an account of the process of discovery of universal gravitation. Preceding the passage above, he wrote:

A stone falls to the ground. Increase the stone's distance from the earth, and it would go against the grain to expect that this continuous increase would lead to some discontinuity. Even at lunar distance the stone will not suddenly lose its tendency to fall. Moreover, big stones fall like small ones: the moon tends to fall to the earth. Our ideas would lose the requisite determination if one body were attracted to the other but not the reverse, thus the attraction is mutual and remains so with unequal bodies, for *the cases merge into one another continuously.* Not only logical elements are at play here: logically, discontinuities are quite conceivable, but it is highly improbable that their existence would not have betrayed itself by some experience. Besides, we prefer the point of view that causes less mental exertion, so long as it is compatible with experience.

(Mach, 1897, pp. 138–139, emphasis added)

By 'logical,' Mach means conceptual, and by 'continuous,' he means fluid and extendable. The fall's distance and the stones' size are the experimental conditions, which are continuously varied in the physicist's mind and eventually stretched to the celestial scale. Reciprocally, the concept of a celestial body, such as the earth and the moon, becomes interchangeable with the concept of a stone, and quite unequal stones can then become mutually attracted. The scope of ideas (expectations) tied to the conditions of the fall of stones is extended simultaneously to the conditions themselves. The cases merge into one another continuously: a conceptual integration is established, connecting near-earth bodies to celestial bodies under a unified physical concept.

In the above example, as in most of Mach's examples, the experimental conditions comprise physical quantities, which makes 'continuous' variation coincide with spanning a real-valued domain. However, a close reading of Mach's entire argument, developed in fourteen numbered analytical steps, suggests that his conception of 'conditions' and their variation, 'continuously if possible,' is broad rather than narrow. That is, although Mach mostly referred to quantitative ideas, he meant the physicist's conceptual representation of sense experience rather than the instantiation of a mathematical model with numerical initial and boundary conditions on physical quantities such as distances, angles, and particle densities. This will be illustrated in what follows.

Mach resumed his account of universal gravitation as a remarkable conceptual integration achieved using the method of continuous variation. He referred to Galileo as a master of this kind of thought experiment and discussed three of his thought experiments, including the one on free-falling bodies:

> If a body of greater weight had the property of falling faster, a combination between a light and a heavy body, would, though heavier still, have to fall more slowly because retarded by the lighter component. The assumed rule is thus untenable because self-contradictory.[104]
>
> (Mach, 1897, p. 139)

Note that properties such as having a 'greater weight,' 'falling faster' and 'fall more slowly' correspond to what Mach calls 'quantitative ideas:'

> Planned quantitative experiment yields many details, but our *quantitative ideas* educated by experiment gain their surest support if we relate them to [unintentionally and instinctively gained] raw experiences. Thus, Stevin adapts his *quantitative ideas* about inclined planes to that experience about the gravity of bodies by means of exemplary thought experiments, and Galileo does likewise with [his quantitative] *ideas* concerning free fall.
>
> (Mach, 1897, p. 141, emphasis added)

Also mentioning Stevin's experiments, Mach connected Galileo's experiments on free fall and inclined planes. In either case, Stevin's or Galileo's, Mach suggests, there is no sharp line distinguishing their thought experiments on inclined planes, on the one hand, and those on the gravity of bodies and falling bodies, on the other hand.[105] Although the inclined plane could be seen as a different setting compared to free-falling bodies, Mach notes that in Galileo's thought, they are the same setting continuously varied, which is done by *modification* and *specialization*. By the method of variation, 'the cases merge into one another continuously,' that is, they are conceptually integrated.

We may now proceed to see how this works in Turing's imitation tests.

5.4.2   *Turing's use of the method of continuous variation*

A reconstruction of Turing's imitation tests as an application of Mach's method of the variation of conditions (continuously, if possible) must show how, in Turing's perspective, the various imitation tests merge into one another continuously. First, it is necessary to introduce Turing's view of creation and evolution, particularly his view of humans, animals, and living beings as machines. Turing's view builds, among other sources, on the organic-machine metaphors of Edwin T. Brewster's *Natural Wonders Every Child Should Know* (1912), which he read in childhood (Hodges, 1983, p. 11).

In (1948), Turing dedicated a section (§3) of his text to describe 'Varieties of machinery.' He observed that '[a]ll machinery can be regarded as continuous, but when it is possible to regard it as discrete it is usually best to do so' (p. 412). A brain, he noted, 'is probably' a 'continuous controlling' machine, but given the digital nature of neural impulses, it 'is very similar to much discrete machinery' (p. 412). Defining the possible states of a machine as a discrete set instead of a continuous set can be convenient for controlling purposes since a 'reasonably accurate knowledge of the state at one moment yields reasonably accurate knowledge any number of steps later' (1950, p. 440). In another section (§6), 'Man as Machine,' Turing construed the differences between 'man' and human-made 'machine' in terms of a continuum:

> A great positive reason for believing in the possibility of making thinking machinery is the fact that it is possible to make machinery to imitate any small part of a man.
>
> (Turing, 1948, p. 420)

Because 'any small part' could be imitated, he imagined:

> One way of setting about our task of building a 'thinking machine' would be to take a man as a whole and to try to replace all the parts of him by machinery. He would include television cameras, microphones, loudspeakers, wheels and 'handling servo-mechanisms' as well as some sort of 'electronic brain'.
>
> (Turing, 1948, p. 420)

He dismissed such a method as 'altogether too slow and impracticable.'

Turing viewed human intelligence in continuity with animal intelligence, as indicated by his formulation of the 'Heads in the Sand' objection to the possibility of machine intelligence: 'We like to believe that Man is in some subtle way superior to the rest of creation' (p. 444). This was an elaboration of objection '(a)' from (1948), which referred to an 'unwillingness to admit the possibility that mankind can have any rivals in intellectual power' (p. 410). Turing, a reader of Samuel Butler,[106] considered human-made machines as a species. In *ca.* mid-1951, he started new foundational research on the genesis and development of organic forms.[107] (In terms of today's concepts of hardware and software, one could say that he started research on organic hardware formation.[108])

For Turing, the differences in intelligence power among all species, human-made machines included, were contingent products of evolution. Until 1950, however, he addressed the problem of intelligence in terms of programming digital computers, which was the technology nearly available for use at the time.[109] Along these lines, Turing suggested machine intelligence could be achieved by making a learning program to simulate a child's mind and subjecting it to 'an appropriate course of education,' in analogy with evolution:

> There is an obvious connection between this process and evolution, by the identifications
>
> Structure of the child machine = Hereditary material
> Changes of the child machine = Mutations
> Natural selection = Judgment of the experimenter
>
> One may hope, however, that this process will be more expeditious than evolution. The survival of the fittest is a slow method for measuring advantages. The experimenter, by the exercise of intelligence, should be able to speed it up.
>
> (Turing, 1950, p. 456).

Turing believed machine intelligence could be progressively developed and eventually achieved by subjecting machines to artificial evolution. The development of the machine's intelligence would depend on the experimenter's intelligence. Although he saw species evolution, whether natural or artificial, physical or cultural, in continuity (i.e., conceptually integrated), his imitation tests presented in (1950) considered hardware fixed (§3) to focus on software instead (§7). Thus by machine evolution, in 1950, he meant cultural evolution.

This clears the ground for shedding light on Turing's use of the method of continuous variation in the design of his imitation tests. There is a core experimental setup based on players A, B, and C and their goals in the imitation game. In Turing's view, all the players are machines in either organic or inorganic, discrete or continuous controlling form. Other types and subtypes apply: woman and man are subtypes of human, which is a subtype of organic and continuous machine. At the same time, differential analyzer and digital computer with its associated learning program are subtypes, respectively, of continuous and discrete non-organic machine. The fundamental question Turing asks (question Q^*) is whether the intellectual and cultural performances associated with the types, namely their related stereotypes, could be imitated, thus empirically showing that the types can be softly transposed.[110] Note that for any arbitrarily chosen type, say, a 'woman,' further specific subtypes can be continuously conceived and considered as varied conditions of the imitation game: women having property p, women having property $p' \subset p$, and so on. Further, for any two arbitrarily chosen types (say, a 'machine' and a 'man'), a new type can be conceived, whether as a specialization or a modification (cf. Turing's thought experimentation on imitating 'any small part of a man'). The existence of such an evolving continuum of levels and types relates to the fact that concepts are fluid entities.[111] This analysis shows how Mach's characterization of the method of continuous variation applies to Turing's imitation tests, or how, in Mach's sense, Turing's variation of conditions aims to make 'the cases merge into one another continuously.'

Further boundary conditions can be varied (continuously if possible): the game's duration, the number of trials, B's actual presence in the game,[112] and the machine's hardware and software capacities. The question across the various versions of the game can be posed this way: how does C's perception of A's performance change as the game's conditions are (continuously) varied? Will it change if gendered verbal behavior, as a subtype of human verbal behavior, is required? Will it change if the game's duration is reduced? Will it change if the machine's hardware is increased and/or its learning program is modified? For Turing, there is no conceptual discontinuity at all among the various conditions that can be chosen for instantiating his thought experiment.

Mach emphasized that 'the basic experimental method of variation' is found within 'man' himself, who collects experiences 'by observing changes,' above all the changes 'he can influence through his own intervention and deliberate movements.' Mach described the playful, instinctive experiments of a child, such as being surprised by their mirror image or shadow in sunlight, and added:

> If the adult temporarily loses these treasures so that he must as it were discover them afresh, the explanation is that his social upbringing narrows his circle of interests and confines him to it while at the same time

he acquires a large number of ready opinions, not to say prejudices, that he supposes not to be in need of examination.

(Mach, 1897, p. 134)

Clearly, Turing did not lose these treasures. Based on Mach's analysis, the fact that Turing's thought experiment involves cultural issues does not make it unscientific. For Mach, '[t]here is no sharp dividing line between instinctive and thought-guided experiments' (p. 134). This is also in line with Floyd's sensible observation (2017) that Turing used common sense as a scientific tool.

At least rhetorically, Turing did not consider that his thought experiment dispensed with physical experiment. He stated that '[t]he only really satisfactory support that can be given' for his positive answer to question Q'''' 'will be that provided by waiting for the end of the century and then doing the experiment described' in the question (1950, p. 455). Once again, this is consistent with Mach's analysis:

If a thought experiment is without definite issue, that is when the idea of certain conditions leads to no certain and unambiguous expectation of a result, we tend to turn to guessing, at any rate for the period between thought and physical experiment, that is we tentatively assume an approximately sufficient condition for a result. This guessing is not unscientific, but a natural process that can be illustrated by historical examples.

(Mach, 1897, p. 141)

Mach further noted: 'The method of letting people guess the outcome of an experimental arrangement has didactic value too' (p. 142).

In light of Mach's analysis, Turing's exposition of his various imitation tests should not be confused with loose rhetoric. Rather than being sloppy, the presentation of his thought experiments can now be understood as methodical. The various questions that Turing asked offered an empirical basis for discussing the original question (can machines think?) under varied limiting conditions. The design of his imitation game was deliberately flexible to address conceptual problems. This observation liberates AI scientists from Turing's specific rhetoric to design, even if Turing-inspired, meaningful, practical experiments. As Mach emphasized, 'thought experiment often precedes and prepares physical experiments' (Mach, 1897, p. 136).

We may now proceed to gain further depth into Turing's uses of his imitation tests and examine what specific conceptual problems they address.

5.5   Turing's critical use of his test

As Chapter 4 has shown, and as is often the case with thought experiments, Turing proposed his test in the context of an intellectual controversy. The significance of the newly existing digital computers was under dispute in postwar England. In 1949, Turing was exposed to strong reactions against his view that machines can think. Although these are described in Chapter 4, it may be worth recalling their general character here.

Fellow of the Royal Society (FRS) and then professor of neurosurgery at the University of Manchester, Geoffrey Jefferson (1886–1961) became Turing's primary intellectual opponent. In his Lister Oration (1949a), Jefferson presented a reductionist view of intelligence, characterized as an emergent property of the animal nervous system. The nervous impulse, he argued, is not a purely electrical phenomenon but also a chemical one that depends on the continuity of specific physical quantities. Further, Jefferson himself used a thought experiment to suggest that gendered behavior is causally related to the physiology of sex hormones (p. 1107). Jefferson's critique of the possibility of machine intelligence was so powerful and comprehensive that it subsumed the objections of other thinkers. For instance, he posited that 'it is cogent argument against the machine that it can answer only problems given to it, and furthermore, that the method it employs is one prearranged by its operator' (p. 1109). This objection was originally championed by Douglas Hartree (1897–1958), FRS and then professor of mathematical physics at the University of Cambridge.[113] Moreover, Jefferson cited René Descartes (p. 1106) and suggested that speech is the distinguishing mark of human intelligence compared to other kinds of animal intelligence (pp. 1109–1110). This also covers the objection formulated by Michael Polanyi (1913–1976), FRS and then professor of social studies at the University of Manchester. Polanyi had presented to Turing a Gödelian argument (cf. Blum, 2010), which later developed into Polanyi's general theory of knowledge (1958). Essentially, according to it, humans can solve problems that machines cannot. Turing was, until then, using the game of chess as a testbed for machine intelligence.[114] However, Polanyi dismissed it as unimpressive (1958): 'A routine game of chess can be played automatically by a machine, and indeed, all arts can be performed automatically to the extent to which the rules of the art can be specified' (p. 261). Jefferson's appeal to speech as the hallmark of human intelligence subsumed Polanyi's argument.

It will be shown that Turing's thought experiment attacks those opposing theories of human intelligence through its varied design. It exemplifies what Popper called the critical use of thought experiments. Moreover, it does so by satisfying Popper's methodological rule for 'the use of imaginary experiments in critical argumentation' (1959), which is to say that *'the idealizations made must be concessions to the opponent, or at least acceptable to the opponent'* (p. 466, no emphasis added).

5.5.1   *The function of the machine-machine variant of the imitation game*

In his Lister Oration (1949a), Jefferson argued that the physiology of the nervous system is based on continuous physical quantities. Therefore, it would be incommensurable with the activity of a digital computer, which, as Turing himself explained, is a discrete-state machine. This is a core element of Jefferson's argument. According to it, thinking is an emergent property that belongs exclusively to the animal nervous system. Therefore, it could not be reproduced by computing machines.

Turing executed the critical use of his test against Jefferson's argument, which he formulated as 'the argument from continuity in the nervous system' (1950, pp. 451–452). He acknowledged that '[t]he nervous system is certainly not a discrete-state machine,' for '[a] small error in the information about the size of a nervous impulse impinging on a neuron, may make a large difference to the size of the outgoing impulse.' However, Turing pondered that this does not mean that a discrete-state system cannot mimic the behavior of the nervous system. He argued that the imitation game neutralizes such a structural difference. He presented an example in which player C asks the other players — A is a digital computer and B is a differential analyzer (a simpler continuous system) — to give the value of a transcendental number such as π. The digital computer could imitate the differential analyzer by choosing at random from a probability distribution between values that approximate the correct answer (say, 3.1416). More generally, the discrete-state machine can use any technique to approximate the continuous-state machine's behavior, and yet an external observer (the interrogator) may not be able to distinguish which is which.

Turing used his test to criticize the argument that a digital computer, as a discrete system, could not imitate human thinking, which is produced by the (continuous) nervous system.

5.5.2   *The functions of player B and the man-woman variant of the imitation game*

Wolfe Mays (1912–2005), who was a contemporary of Turing at the University of Manchester and another opponent of his views (Mays, 2001), guessed that a specific source for Turing's imitation game was Twenty Questions (1952, p. 148), a radio parlor game that Turing had made casual reference to in his own writing (1950, p. 457). Hodges (1983) noted that Turing played the latter with friends during a summer holiday and even 'developed a theory of how to choose the next question so as to maximise the expected weight of evidence of the answer' (p. 389). In the game, the players must identify an entity by asking up to twenty yes-no questions. The only clue that can be provided is whether the item was of animal, vegetable, or mineral nature, which highlights the game's focus on ontological categories. Sterrett found (2020) that since the early 1950s, there have been TV shows whose structure is even more similar to Turing's imitation game. Inspired by parlor games, the Turing test suits Mach's point that thought experiments are sourced in quasi-sensory information such as combinations of memories of sense elements (1897, p. 137).

But why did Turing specifically address the problem of sexual guessing? Sterrett (2000) argued that player A needs to think reflectively to avoid giving ingrained responses that would reveal their true kind, and gender, depending on the culture and how one is raised, can be such an ingrained characteristic of an individual. She remarked that 'cross-gendering is not essential to the test; some other aspect of human life might well serve in constructing a test that requires such self-conscious critique of one's ingrained responses' (p. 470). This insight introduced by Sterrett's analysis captures in a fundamental way the intellectual skill required from player A across the various conditions presented by Turing's imitation tests: to turn into what he/it is not. Nevertheless, this question remains: among other possible properties that Turing could have chosen (as Aristotelian differentia for the human genus) in his use of the method of variation, why did he choose gender in particular?

The capability to think through gender had a specific role in Turing's critical use of his thought experiment. Jefferson presented a critique of the artificial behavior of 'modern automata' (1949a, p. 1107). He referred to the then famous electromechanical tortoises of the cybernetician Grey Walter and, in doing so, offered Turing an imaginary experiment:

> [...It] should be possible to construct a simple animal such as a tortoise (as Grey Walter ingeniously proposed) that would show by its movements that it disliked bright lights, cold, and damp, and be apparently frightened by loud noises, moving towards or away from such stimuli as its receptors were capable of responding to. In a favourable situation the behaviour of such a toy could appear to be very lifelike — so much so that a good demonstrator might cause the credulous to exclaim 'This is indeed a tortoise.' I imagine, however, that another tortoise would quickly find it a puzzling companion and a disappointing mate.
>
> (Jefferson, 1949a, p. 1107)

It can be argued that a key function of Turing's 1950 imitation tests is to criticize this thought experiment on automata and gender, which they partly reconstruct. Jefferson brought forward the image of a genuine individual of a kind, which is placed side by side with the artificial one so that the latter's artificiality is emphasized. The function of the genuine individual is to reveal the artificiality of the imposter. That explains Turing's introduction of a control player (B), which only appears as a structural element in the 1950 variants of Turing's imitation tests. In the (1948; 1951a; 1952) tests, the machine plays directly against the judge with no control player around. With Popper's rule in mind, the control player can be explained as a concession to Jefferson.

Jefferson referred to 'sex hormones' as a distinctive feature of the intelligent behavior of 'animals' and 'men,' as opposed to 'modern automata' (1949a, p. 1107). He remarked that 'neither animals nor men can be explained by studying nervous mechanics in isolation, so complicated are they by endocrines, so coloured is thought by emotion.' He then added: '[s]ex hormones introduce peculiarities of behaviour often as inexplicable as they are impressive' (p. 1107). In effect, Jefferson suggested that machines could not exhibit enough peculiarities of behavior to imitate the actions of animals or 'men' because they are not moved by sex hormones. A machine would give itself away and be found to be 'a puzzling companion and a disappointing mate.' In a further passage,[115] Jefferson stated that he would not agree that 'machine equals brain' until a machine could, among other things, 'be warmed by flattery' and 'be charmed by sex' (p. 1110).

In summary, Jefferson substantiated his argument that human intelligence is an exclusive product of the physiology of the animal nervous system with the thesis that gendered behavior is a causal product of male and female sex hormones. For Turing to meet Jefferson's challenge and conceive a machine that could be convincingly human-like, as opposed to a puzzling companion and a disappointing mate, it would have to be able to learn and successfully imitate gender. The function of Turing's man-woman control variant of the game was to establish, through the simple common sense of a parlor game, that gender stereotypes can be learned and imitated despite the players' physiological differences. Turing thus established from the start of his 1950 text (Section 5.3.1) that question $Q^\star$ (Section 5.3.2) can be meaningful from a logical point of view (it is not a conceptual paradox) and, therefore, open for empirical study. In other words, rather than serving as a scoring protocol to Q''', Q' serves a rhetorical purpose within the critical function of the Turing test.

Further, the man-woman game tries to expose the existence of a conceptual paradox within Jefferson's theory that physical kind determines logical kind — if a man can imitate intellectual stereotypes associated with a woman despite their physical differences, why could a machine not imitate a woman, a man, or, more broadly, a human? That satisfies Kuhn's characterization of the function of thought experiments (1964), for Turing proposed a conceptual change on the traditional concepts of machine and intelligence at the time,[116] which Jefferson had articulated in scholarly form using his background in neurophysiology.

The machine-woman case variant of the game reinstates the question of the learning and imitation of gender stereotypes as a challenging special case of question Q^*.

5.5.3   The function of conversation as the intelligence task addressed by the imitation game

Since his wartime service from 1941 to late 1949, Turing considered the game of chess as his chosen intelligence task to illustrate, develop and test machine intelligence. In 1948, he discussed a tradeoff between convenient and impressive intellectual fields for exploring machine intelligence. Regarding language, and having discussed 'various games, e.g., chess,' Turing wrote (1948): 'Of the above possible fields the learning of languages would be the most impressive, since it is the most human of these activities' (p. 421).[117] However, he pondered, that field seems 'to depend rather too much on sense organs and locomotion to be feasible.' In the end, he kept his choice for chess and described a chess-based imitation game (p. 431).

Eventually, as mentioned, Turing's use of chess to test for machine intelligence was directly challenged by Polanyi and indirectly challenged by Jefferson. From 1949 to 1950, Turing changed his option and built his thought experiment in the form of a conversation game. Unlike chess, which is governed by definite rules, good performance in conversation cannot be easily specified. Therefore, Turing's 1950 choice for 'the learning of languages' as the intellectual field addressed in his test can be best understood as yet another concession to Jefferson and, in this case, to Polanyi as well.

Now, note that the machine-man case variant of the game is designed to test the machine's capability of *language learning*, which is Turing's specific uptake of the required skill (language use and understanding). If Turing's various imitation tests are understood as part of his continuously varied thought experiment (Section 5.3), the exegetical problem of whether Turing meant masculine generics in the machine-man game vanishes. That is because gendered language learning, as a challenging special case of natural language learning, had already been implied as a required skill by the machine-woman game.

5.6   Turing's heuristic use of his test

Turing considered his imitation game as a means to distinguish true language learning from parrot-fashion learning. He addressed this issue also in his response to Jefferson's demand that a thinking machine should be able to create a sonnet on its own (1949a, p. 1110). Turing thus presented this example of an exchange between his imaginary machine and player C, the human interrogator, who questions it about a sonnet that it has written:

> Probably he [Jefferson] would be quite willing to accept the imitation game as a test. The game (with the player B omitted) is frequently used in practice under the name of *viva voce* to discover whether some one really understands something or has 'learnt it parrot fashion'. Let us listen in to a part of such a *viva voce*:
>
> Interrogator: In the first line of your sonnet which reads 'Shall I compare thee to a summer's day', would not 'a spring day' do as well or better?
> Witness: It wouldn't scan.
> Interrogator: How about 'a winter's day'. That would scan all right.
> Witness: Yes, but nobody wants to be compared to a winter's day.
> Interrogator: Would you say Mr. Pickwick reminded you of Christmas?
> Witness: In a way.
> Interrogator: Yet Christmas is a winter's day, and I do not think Mr. Pickwick would mind the comparison.
> Witness: I don't think you're serious. By a winter's day one means a typical winter's day, rather than a special one like Christmas.
>
> And so on. What would Professor Jefferson say if the sonnet-writing machine was able to answer like this in the *viva voce*? I do not know whether he would regard the machine as 'merely artificially signalling' these answers, but if the answers were as satisfactory and sustained as in the above passage I do not think he would describe it as 'an easy contrivance'.
>
> (Turing, 1950, pp. 446–447)

To understand the heuristic function of Turing's test in the Popperian sense, it is important to emphasize what Turing's imaginary sonnet-writing machine *illustrates* (Section 5.6.1) and what it *suggests* (Section 5.6.2).

5.6.1   *Turing's test illustrates a property of the phenomenon of intelligence*

Turing presented a standard of intelligent behavior that he thought could be produced by a machine. He believed that the imaginary machine's performance was so 'satisfactory and sustained' that it would stress Jefferson's aprioristic claim that, whatever a machine could do, it would be nothing but a result of shallow symbol manipulation. The practical Turing tests (Section 5.1.1) have shown that Jefferson's point still stands. Whether Turing may have underestimated the power of modern mechanical parrots will be discussed later (Section 5.6.2).

In any case, it is worth noting Turing's manifest uncertainty on how the machine's performance, which he took to be suggestive of true language understanding, would be perceived by Jefferson (perhaps as a mere artifice). Turing had noted (1948) that some of the objections to the possibility of machine intelligence were 'purely emotional' (p. 411); therefore, the justification of an intelligence claim could not rest on logic alone. This is an important point illustrated by the heuristic function of the imitation game. The game encodes Turing's insight that explaining 'the cause and effect' of mechanical intelligence makes it unimpressive and seem 'a sort of unimaginative donkey-work' that is unworthy to be called thinking (1952, p. 500). For that reason, the imitation game has been designed to be a blind experiment centered on behavior rather than on internal states: 'Usually if one maintains that a machine can do one of these things, and describes the kind of method that the machine could use,' Turing remarked in (1950), 'one will not make much of an impression' (pp. 449–450). It was instead 'the actual production of the machines,' Turing had guessed in (1948), that 'would probably have some effect' (p. 411). This explains Turing's use of an imaginary (machine) experiment at a time when he was still waiting for the Manchester Automatic Digital Machine to be available for his first preliminary experiments (Lavington, 2012, p. 99).

Diane Proudfoot (2013) identified in two of Turing's works (1948; 1952) his view that the perception of intelligence is emotional,[118] which she developed into a response-dependence theory of intelligence. This means that a machine can be said to be intelligent if it appears intelligent to 'a normal subject' in certain 'specified conditions' of observation (Proudfoot, 2013, p. 404). In fact, Proudfoot argued (2017b), *'the Turing test does not test machine behaviour'* (p. 303, no emphasis added). 'Instead,' she wrote, 'it tests the observer's reaction to the machine.' This pushes the Turing test closer to psychometrics and farther from AI. Response dependence can be illustrated through other secondary-quality concepts. For example, a color can be perceived similarly by people who are not colorblind in adequate lighting conditions. This, of course, does not preclude a physics of color, which reifies color as a (response-independent) primary quality concept. However, Proudfoot commits to a notion of 'global response-dependence' (Pettit, 1991, p. 588). This imputes to Turing the view that intelligence is a socially constructed concept whose verifiability rests on the intersubjective judgment of human interrogators. If an unintelligent chatbot fools humans under the specified conditions, the chatbot can be claimed intelligent. Proudfoot takes 'the concept of colour' as being 'very different from the concept of electromagnetic radiation, even though electromagnetic radiation is the physical basis of colour.' 'Likewise,' Proudfoot concludes (2017b), 'if intelligence is a response-dependent concept, the concept of intelligence is very different from the concept of computation, *even if* brain processes (implementing computations) form the physical basis of "thinking" behaviour' (p. 305, no emphasis added). Essentially, Proudfoot commits to anti-physicalism: she rejects the reification of the physical concepts of color and intelligence as primary-quality concepts.

Turing, however, did refer to intelligence as a dispositional physical property grounded in material computational power. In (1948), he referred to the 'intellectual power' of humankind and other animal species (p. 410) and the 'intellectual power' that the 'isolated man' cannot develop given his limited possibilities for learning (p. 431).[119] In (1950), he referred to 'the power of thinking' (p. 444); and in (1952), he said that 'an intelligent human mind' could learn how to learn (p. 497). Turing's physical concept of intelligence and its connection to the Turing test has been explained by his colleague Donald Michie as follows:

> Turing's belief about intelligence was that the *PROPENSITY* is *INNATE*, but the *ACTUALITY* has to be *BUILT*. For him the crux was the brain's ability to make sense of its inputs, that is to *understand* them. And how would we tell whether we had succeeded? To assess degrees of machine understanding he was later to propose what is celebrated today as the *Turing Test*.
>
> (Michie, 2002, no emphasis added)

This oral source suggests that Turing did consider intelligence a physical concept and his test a sort of experiment for machine intelligence.

Nevertheless, Turing's experience with Jefferson and others showed that actual intelligence (on the computer, as in the brain) was not enough to *justify* a machine intelligence claim. Especially in the early 1950s, when the traditional concept of intelligence was tied to humans, justifying machine intelligence in terms of inner computational structures would make a circular argument. Instead, machine intelligence had to be demonstrated by addressing language use and understanding—a skill that indisputably belonged to human intelligence—, so that it could be *perceived*. Illustrating this is the first part of the heuristic function of the Turing test.

Now, if Turing relied on his test to assess machine understanding, did he overestimate the capacity of human interrogators to unmask mechanical parrots?

5.6.2 *Turing's test suggests a hypothesis on machine learning*

Human-like chatbots can be based on a combination of psychological tricks and ad-hoc schemes to store and retrieve human-built, semi-structured content pulled from the Internet.[120] From a conceptual point of view, machines of this kind can be understood as sophisticated mechanical parrots. For a related example, Sterrett (2020) described how IBM researchers built the unintelligent Watson system to outstrip humans in the popular *Jeopardy!* game by using Internet-based content and exploiting the *a priori* known structure of the game (pp. 473–474).

Turing reprobated the use of 'the man inside the machine' stratagems that characterizes the top-ranked machines that competed in practical Turing tests thus far. In (1951b), he posited that the machine learning processes that he envisioned 'could probably be hastened by a suitable selection of the experiences to which [the machine] was subjected' (p. 473). 'But here,' Turing warned, 'we have to be careful.' 'It would be quite easy,' he continued, 'to arrange the experiences in such a way that they automatically caused the structure of the machine to build up into a previously intended form.' This, he adverted, 'would obviously be a gross form of cheating, almost on a par with having a man inside the machine.' In other words, Turing ruled out from his test machines that are specially conditioned to pass it, just like IBM Watson was specially conditioned for *Jeopardy!*.

For Turing, of course, a machine 'having a man inside' could never be an existence proof of machine intelligence. On the other hand, mechanical parrots disregarded, he considered that the conversation performance of his imaginary sonnet-writing machine could hardly have been produced unless it had truly learned about British Christmas traditions, characters in Charles Dickens' novel, the use of sarcasm, and so on. For Turing, such a performance would be best explained by assuming a true learning and understanding of the English language and the related culture, just as is assumed in viva voce examinations.

Yet, how many examinations should be enough for an existence proof? Turing said:

> It is clearly possible to produce a machine which would give a very good account of itself for any range of tests, if the machine were made sufficiently elaborate. However, this again would hardly be considered an adequate proof. Such a machine would give itself away by making the same sort of mistake over and over again, and being quite unable to correct itself, or to be corrected by argument from outside. If the machine were able in some way to 'learn by experience' it would be much more impressive.
>
> (Turing, 1951b, p. 473)

This passage could be read as supporting the positive answer to the Turing Test Dilemma: Turing believed that unrestricted tests would eventually unmask elaborate yet unintelligent machines. However, is running repeated unrestricted tests on unintelligent machines valuable for AI? Shieber (1994a) noted that unrestricted Turing tests—precisely for being unrestricted—could not support scientific progress in AI. Therefore, seeing the Turing test as a practical experiment reduces its value to its confirmatory power. However, this pushes the test nearer to the psychometric issues related to the judgment of average human interrogators and farther from AI research.

Now, the interpretation of the Turing test as a thought experiment in the modern scientific tradition presents another reading of the above passage, which observes what Turing *suggested*: even elaborate machines could not qualify as 'an adequate proof' of human-level machine intelligence if they could not learn from experience to correct themselves or be corrected without reboots. In fact, Turing held a specific view of what an existence proof would be (1950, pp. 455–459): to raise a simple learning machine through an adapted process of language and culture education that should be analogous to the one that a human child goes through, until it could, without reboots or special coaching, play the imitation game well.[121] The second part of the heuristic function of the Turing test is to suggest that this is possible,[122] as developed next.

Turing's concern was not the design of a practical experiment whose confirmatory power would be robust against *false* positives. It was instead the proposal of an empirical criterion for justifying an existence proof of machine intelligence in the presence of *true* positives. As Shieber observed more recently (2016), the Turing test 'works exceptionally well as a *conceptual* sufficient condition for attributing intelligence to a machine, which was, after all, its original purpose' (p. 95, emphasis added).

Yet, why would the playful imitation game be such an adequate proof of the revolutionary possibility of intelligent machinery? If Michie was correct that the test was meant to assess 'machine understanding,' how can Turing's focus on deception be explained?[123]

First, it is worth recalling the question Q^* that can be generalized from Turing's presentation of his test (Section 5.3): could player A imitate intellectual stereotypes associated with player B's type successfully (well enough to deceive player C), despite the physical differences between A's and B's types?

In fact, given that the perception of intelligence involves emotion (Section 5.6.1), deception, or the capability to manipulate the states of mind of another agent, must be addressed as an intrinsic meta-task in any experiment related to Q^*. The Turing test, therefore, prepares for related practical experiments addressing deception in AI. As Mach remarked, 'thought experiment often precedes and prepares physical experiments' (1897, p. 136).

Proudfoot (2011) has urged AI scientists to acknowledge the value of the Turing test as a practical experiment, and her position must face the second horn of the Turing Test Dilemma. However, this chapter's reconstruction of the Turing test as a thought experiment preserves a deflationary view of her argument, which shows how the Turing test introduced the idea that deception can be and should be controlled for in AI experiments.

Sterrett (2020) contributed an analysis that does justice to Turing's distinction between, on the one hand, the perception of intelligence as grounded in deception in the context of a game and, on other hand, intelligence itself as grounded in learning. Sterrett explained how the Turing test addresses deception through a comparative analysis of popular parlor games. 'The game context,' she remarked, 'provides means to hone in on the part of language performances that have to do with being reflective and resourceful, i.e., not "machine-like"' (p. 471). The intellectual abilities required by impersonation, Sterrett highlighted from a passage in Ryle's *The Concept of Mind* (1949, p. 33), are perhaps most clearly pronounced in the performance of a clown. Observing Turing's background in espionage, the performance of an intelligence agent may also be considered. Deception can be hard even for a sophisticated mechanical parrot to simulate if not resorting to special coaching by the human programmer 'inside' it.

The distinction between true machine education and special coaching appears in Turing's guidelines on how the machine should be programmed. He addressed that distinction through his heuristic execution of the imitation game. He observed that the imitation of *human fallibility* is necessary for deceiving a human observer. He illustrated human fallibility, first, in the form of incapacity for sonnet-writing, and second, in the form of an arithmetic mistake:

Q Please write me a sonnet on the subject of the Forth Bridge.
A Count me out on this one. I never could write poetry.
Q Add 34957 to 70764
A (Pause about 30 seconds and then give as answer) 105621.

(Turing, 1950, p. 434)

Now, the key point here is to note how human fallibility appears in Turing's vision of AI:

Another important result of preparing our machine for its part in the imitation game by a process of teaching and learning is that 'human

> fallibility' is likely to be [mimicked] in a rather natural way, *i.e.*, without special 'coaching' … Processes that are learnt do not produce a hundred per cent. certainty of result; if they did they could not be unlearnt.
>
> (Turing, 1950, p. 459, no emphasis added)

In effect, the coherence of the Turing test rests in that the machine's capability to deceive the human interrogator about its true kind must be a corollary of its own learning from experience.

The second part of the heuristic function of the Turing test is to suggest the hypothesis that a learning machine may be created simple and educated naturally, without reboots or special coaching, to play the imitation game well.

5.7   Conclusion

This chapter has shown that the Turing test can be best understood as a thought experiment in the modern scientific tradition. First, it has shown that underlying Turing's 1950 presentation of various imitation tests (Section 5.3.1), there is a rich methodological structure (Section 5.3.2), which conforms to what Mach characterized as the basic method of thought experiments, consisting of a continuous variation of experimental conditions (Section 5.4).

Second, this chapter has presented a reconstruction of Turing's thought experiment that satisfies Popper's conception of the critical and the heuristic uses of imaginary experiments. That reconstruction has emphasized how the Turing test increases the understanding of the question 'can machines think?' and prepares for related practical experiments. This provides a rapprochement to the conflicting views on the value of the Turing test for AI and can ultimately put an end to the Turing Test Dilemma as a two-horned issue.

Specifically, this chapter has shown how Turing's methodic variation of his test design consists of a critical use of the test against the view that physical kind determines logical kind (Section 5.5). The various forms of the test, rather than being a result of imprecisions and bad design choices, as suggested in the secondary literature, can be seen instead as concessions to Turing's intellectual opponents. This conforms to Popper's rule for using imaginary experiments in critical argumentation and puts an end to the first horn of the dilemma. Turing's imitation tests addressed the following opposing theories of intelligence presented to Turing:

1   Human-level intelligence is an exclusive product of the physiology of the animal nervous system, and gendered behavior is a causal product of male and female sex hormones (Jefferson).

2 A machine can only do what it has been instructed to do (Lovelace-Hartree).

3 A given art can be performed automatically only to the extent that its rules can be specified, as in the game of chess (Polanyi).

In particular, this chapter has shown that, seeking conceptual change, Turing used his imitation game to reveal a paradox in the theory of intelligence presented by Jefferson, which tied logical kind to physical kind. This satisfies Kuhn's characterization of the function of thought experiments.

Further, this chapter has reconstructed Turing's heuristic use of his test (Section 5.6), showing that the test illustrates the emotional nature of the perception of intelligence. This explains why the practical value of the test necessarily depends on the judgment of (average) human interrogators. However, Turing also used his test to suggest the hypothesis that a learning machine may be created simple and educated naturally, without reboots or special coaching, to play the imitation game well. This explains why running practical Turing tests on machines that have been specially coached to pass it is misguided. The focus of Turing's proposal was to provide both an empirical criterion to justify an existence proof of machine intelligence and a research strategy for fulfilling that criterion. The reconstruction of Turing's heuristic use of his test puts an end to the second horn of the dilemma.

Mach (1897) observed that thought experiments based on continuous variation 'undoubtedly have led to enormous changes in our thinking and to an opening up of most important new paths of enquiry' (p. 138). This is the case with the Turing test.

6   Galilean Resonances

In 1950, Alan Turing proposed his iconic imitation game, calling it a 'test,' an 'experiment' and the 'the only really satisfactory support' for his view that machines can think. Following Turing's rhetoric, the 'Turing test' has been widely received as a kind of crucial experiment to determine machine intelligence. However, in 1951 and 1952, Turing showed a milder attitude towards what he called his 'imitation tests.' In 1948, Turing referred to the persuasive power of 'the actual production of machines' rather than that of a controlled experiment. Observing this, this chapter proposes to distinguish the logical structure from the rhetoric of Turing's argument. It is argued that Turing's proposal of a crucial experiment may have been a concession to meet the standards of his interlocutors more than his own, while his construction of machine intelligence rather reveals a method of successive idealizations and exploratory experiments. A parallel is drawn with Galileo's construction of idealized fall in a void and the historiographical controversies over the role of experiment in Galilean science. Turing, like Galileo, relied on certain kinds of experiment, but also on rhetoric and propaganda to inspire further research that could lead to convincing scientific and technological progress.[*]

6.1   Turing's rhetoric of a crucial experiment

Turing opened his seminal paper (1950) by proposing to replace the question 'Can machines think?', which he considered 'too meaningless to deserve discussion' (p. 442), with a new one. The new question would have a 'more accurate form' and would be based on what he called the 'imitation game' and, later in the same text, his 'test' and 'experiment.'[124] Following Turing's rhetoric of experiment, the 'Turing test' has been widely construed in analytic philosophy and cognitive science (Moor, 2003; Shieber, 2004; Copeland and Proudfoot, 2009; Proudfoot, 2013), artificial intelligence (Hayes and Ford, 1995; Proudfoot, 2011), and within Turing scholarship itself (Copeland, 2017; Proudfoot, 2017a), as a kind of crucial experiment to determine the existence of an intelligent machine. Interestingly, Turing offered a testable prediction, which

---

[*]This is chapter is based on Gonçalves, B. (2023). Galilean resonances: the role of experiment in Turing's construction of machine intelligence. *Annals of Science* (forthcoming). doi: 10.1080/00033790.2023.2234912

comes in his 1950 text immediately after his discursive presentation of his test:

> I believe that in about fifty years' time it will be possible to programme computers, with a storage capacity of about 10^9, to make them play the imitation game so well that an average interrogator will not have more than 70 per cent, chance of making the right identification after five minutes of questioning.
>
> (Turing, 1950, p. 442)

This prediction has often been read to mean that if a machine can play the imitation game well enough in five-minute sessions to fool an average interrogator in at least one-third of the sessions, then it passes the Turing test. But this has led to controversy.

Kevin Warwick and Huma Shah (2015) organized a practical implementation of the Turing test based on the above reading of that prediction, which led them to announce a chatbot that would have been 'the first to pass the Turing test.'[125] When questioned about the chatbot, which relied on information-retrieval tricks to deceive observers, Warwick and Shah (2015) insisted that their Turing test experiment was designed 'as set out by Alan Turing.' To refute their Turing test-pass claim, Jack Copeland (2017) argued that Turing's 1950 prediction does not specify the rules of the test, but rather refers to the rate of progress Turing expected towards passing his test (pp. 272–273). Copeland interprets Turing's 1950 prediction with the support of another prediction Turing made less than two years later in a BBC broadcast (1952). When asked how long it would take for a machine 'to stand any chance' in his test 'with no questions barred,' he replied 'at least 100 years, I should say' (p. 495). Now, this interpretation selects one particular projection and focuses attention on it, while there are several others that Turing made rhetorically, in passing, in the context of a debate. In May 1951, for example, he had given yet another estimate: 'I think it is probable for instance that at the end of the century it will be possible to programme a machine to answer questions in such a way that it will be extremely difficult to guess whether the answers are being given by a man or by the machine' (1951a, p. 484). Further, Copeland's interpretation sacrifices consistency within Turing's 1950 paper as a single exegetical unit, for it is in tension with how Turing himself refers to his prediction in his very 1950 paper and the support he expects from the outcome:

> The reader will have anticipated that I have no very convincing arguments of a positive nature to support my views ... If I had I should not have taken such pains to point out the fallacies in contrary views. Such evidence as I have I shall now give. The only really satisfactory support

that can be given for the view expressed at the beginning of §6 [his stated prediction], will be that provided by waiting for the end of the century and then doing the experiment described. But what can we say in the meantime? What steps should be taken now if the experiment is to be successful?

(Turing, 1950, pp. 454–455)

The prediction is twice referred to as 'the experiment,' and it is the 'only really satisfactory support' for his view that machines can think. Turing went on to describe a research strategy for building 'learning machines' that would play the imitation game well, so that the experiment could be 'successful' when performed 'at the end of the century' (1950, pp. 455–459). Turing writes in that passage as if he were a card-carrying empiricist who believed that controlled experiment was the only source of reliable knowledge about the natural world, as if he believed that experiment had a demonstrative purpose and that his question — Can machines think? — should be decided by an experiment.[126]

However, a close reading of Turing's 1950 paper, in conjunction with earlier and later primary sources, can reveal important tensions in interpreting his rhetoric of experiment at face value. In several passages, Turing shows a milder attitude towards the confirmatory power of his 'imitation tests,' and he refers to the demonstrative power of scientific and technological progress rather than that of a controlled experiment. In Turing's major report on 'Intelligent Machinery' (1948) the rhetoric of crucial experiments does not appear. It was rather 'the actual production of the machines' that could 'have some effect' in convincing critics and opponents, because 'the idea of "intelligence" is emotional rather than mathematical' (p. 411). Later in (1950), when Turing formulates his 1950 prediction, he carefully distinguishes his imitation test as but 'the more accurate form of the question.' He writes: 'The original question, "Can machines think?," I believe to be too meaningless to deserve discussion.' 'Nevertheless,' he adds, 'I believe that at the end of the century the use of words and general educated opinion will have altered so much that one will be able to speak of machines thinking without expecting to be contradicted' (p. 442).[127] Furthermore, in (1951a; 1951b) Turing discusses his belief in the possibility of machine intelligence at length, and does not refer to experiments except for a minor mention of his 'viva voce examination' in (1951a, p. 484). In (1952) he reformulates his imitation test into a different scenario, where the machine is interrogated by 'a jury, who should not be expert about machines' (p. 495). Referring to this additional variant of the imitation test, Turing drops the language of associating thinking and intelligence with machines, which he had admitted to be 'heretical' only a few months before on the radio (1951a), and suggests instead that machines passing such a test could be considered 'grade A' machines (Turing et al., 1952). Overall, he never provides detailed settings for any of his various imitation tests as controlled experiments.

6.2   Argument sketch

Observing these tensions, this chapter proposes to distinguish the logical structure of the Turing test argument from its rhetoric. I argue that the idea of a crucial experiment appears in Turing's work as a specific move in 1950, probably as a concession to meet the standards of his interlocutors more than his own. I will start by examining Turing's construction of machine intelligence, in which a method of successive idealizations and exploratory experiments can be observed (Section 6.3). Next, I will propose to read Turing's 1950 text according to a specific logical structure (Section 6.4). The proposed analysis will also relate the main logical step of Turing's 1950 argument to a possible source, namely Bertrand Russell's account of the Socratic dialectic method in his *A History of Western Philosophy*, first published in 1945. I will then draw a parallel with Galileo's construction of idealized fall in a void and the historiographical controversies surrounding his rhetoric of experiment (Section 6.5). Following Galilean studies, I will argue for a distinction between the method Turing actually used and his comments on methodology, since it can be noted that his method was not a direct result of his comments on methodology and vice-versa. Overall, I argue that Turing, comparably with Galileo, relied on certain kinds of experiment, but also on rhetoric and propaganda, apparently recognizing the need to inspire further research that could lead to convincing scientific and technological progress (Section 6.6). The main points are summarized at the end (Section 6.7).

6.3   Turing's construction of machine intelligence

At the outset, it may be worthwhile to distinguish between Turing's construction of machine intelligence and of his imitation tests. On the one hand, his imitation tests can be understood as illustrating that the human perception of intelligence is an emotional phenomenon, and as suggesting a criterion of success and a goal for the study of machine intelligence. On the other hand, his communications and writings show that he developed a theory of mechanical intelligence inspired by human cognitive development and the workings of the human brain, particularly the cerebral cortex of a human child.

I will now examine the role of experimentation in Turing's development of both elements. Essentially, he used a methodology of successive idealizations and exploratory experiments, most of which he described in his major report 'Intelligent Machinery' (1948).

6.3.1   *Exploratory imitation tests on the human perception of machine intelligence*

In 1948, Turing still did not have access to a digital computer, something that he had to wait for until 1951.[128] He thus conceived the notion of a 'paper' machine:

> It is possible to produce the effect of a computing machine by writing down a set of rules of procedure and asking a man to carry them out. Such a combination of a man with written instructions will be called a 'Paper Machine.'
>
> (Turing, 1948, p. 416)

Turing described how he studied the human perception of machine intelligence by means of thought-guided experiments on chess-playing 'paper' machines. Arguing against the idea that machines could not surpass the intelligence of their creators, he reports to have actually experienced the 'sort of situation,' where the machine goes beyond the intelligence of its creator, 'arises in a small degree.' 'Playing [chess] against such a machine,' he wrote, ' gives a definite feeling that one is pitting one's wits against something alive' (p. 412).

The general point that Turing was trying to make, and which he seems to have learned through his exploratory experiments, is that the perception of intelligence has a subjective component. He had just stated that some objections to the possibility of machine intelligence, which he considered 'purely emotional,' could not be 'wholly ignored, because the idea of "intelligence" is itself emotional rather than mathematical' (p. 411). He returned to this important point in a final section of his report (1948), 'Intelligence as an emotional concept,' writing:

> The extent to which we regard something as behaving in an intelligent manner is determined as much by our own state of mind and training as by the properties of the object under consideration. If we are able to explain and predict its behaviour or if there seems to be little underlying plan, we have little temptation to imagine intelligence. With the same object therefore it is possible that one man would consider it as intelligent and another would not; the second man would have found out the rules of its behaviour.
>
> (Turing, 1948, p. 431)

Turing notes that intelligence is perceived by an observer in the behavior of an agent. The observer's perception is influenced by the agent's actual capabilities as well as the observer's own biases. The less the observer knows about the agent's mechanisms, the more (s)he is tempted to 'imagine' intelligence. It is interesting to note that Turing arrived at this insightful observation with the help of an initial experiment with a paper machine, which he describes in the same passage, at the end of his report:

> It is possible to do a little experiment on these lines, even at the present stage of knowledge. It is not difficult to devise a paper machine which will play a not very bad game of chess. Now get three men as subjects for the experiment A, B, C. A and C are to be rather poor chess players, B is the operator who works the paper machine. (In order that he should be able to work it fairly fast it is advisable that he be both mathematician and chess player.) Two rooms are used with some arrangement for communicating moves, and a game is played between C and either A or the paper machine. C may find it quite difficult to tell which he is playing. (This is a rather idealized form of an experiment I have actually done.)
>
> (Turing, 1948, p. 431)

This is evidently a first variant of an imitation test, in this case using chess as opposed to conversation as the intellectual task of the test. Turing even notes that this is 'a rather idealized form of an experiment,' as if aware that his style of reasoning may not meet the highest empiricist standards of what qualifies as a controlled physical experiment.

Turing reiterated his point about the perception of intelligence as an emotional phenomenon in 1950,[129] and 1952.[130]

6.3.2   *Exploratory experiments on child machines and their learning capabilities*

Also in 1948, Turing referred to another kind of initial experiment on machine intelligence to explore the possibility of 'organizing' an initially 'unorganized' machine. The reference for an organized machine was his 1936 'universal' machine, which was an idealization of '[a] man provided with paper, pencil, and rubber, and subject to strict discipline' (1948, p. 416). Turing used this reference to explain the 'unorganized' machine instead as an idealization of the brain cortex of a human child, which can be molded:

> We believe then that there are large parts of the brain, chiefly in the cortex, whose function is largely indeterminate. In the infant these parts do not have much effect: the effect they have is uncoordinated. In the adult they have great and purposive effect: the form of this effect depends on the training in childhood. A large remnant of the random behaviour of infancy remains in the adult. All of this suggests that the cortex of the

infant is an unorganised machine, which can be organised by suitable interfering training.

(Turing, 1948, p. 424)

The 'unorganized' machine differs from the 'universal' machine in its flexibility and structural disposition for learning. It is worth pausing for a clear understanding of Turing's machine conceptions. The unorganized machine, as a working idealization of a human child's cortex, should have the potential to balance 'discipline' with 'initiative:'

> If the untrained infant's mind is to become an intelligent one, it must acquire both discipline and initiative. So far we have been considering only discipline. To convert a brain or machine into a universal machine is the extremest form of discipline. Without something of this kind one cannot set up proper communication. But discipline is certainly not enough in itself to produce intelligence. That which is required in addition we call initiative. This statement will have to serve as a definition. Our task is to discover the nature of this residue as it occurs in man, and to try and copy it in machines.
>
> (Turing, 1948, p. 429)

This passage gives in general lines what Turing explored in his initial experiments on how to develop machine intelligence. The experiments were paper-based and focused on the unorganized machine, while taking the universal machine as a reference for overly rigid but functional behavior. He expected to be able to partially organize the former under the interference of a teacher giving it reward and punishment feedback. If successful, this would be a proof of concept that a machine *could* potentially exhibit functional (and in this case flexible) behavior through learning, without being directly told what to do. In Turing's view, this would make the machine intelligent. Along these lines, he entitled a section (§10) of his 1948 report 'Experiments in organising. Pleasure-pain systems' (p. 424) and thus opened it:

> It is interesting to experiment with unorganised machines admitting definite types of interference and trying to organize them, e.g. to modify them into universal machines ... I have investigated a particular type of pleasure-pain system, which I will now describe.
>
> (Turing, 1948, p. 424–425)

Turing explicitly noted that his experiments were exploratory in nature and their point was to allow him to refine his unorganized machine as an idealization of a human child's cortex:

The actual technique by which the 'organising' of the P-type [pleasure-pain] machine was carried through is perhaps a little disappointing. It is not sufficiently analogous to the kind of process by which a child would really be taught ... I feel that more should be done on these lines. I would like to investigate other types of unorganised machine, and also to try out organising methods that would be more nearly analogous to our 'methods of education.'

(Turing, 1948, p. 428)

Later, in (1950), Turing reiterated the same point about the nature of his paper-based experiments on machine intelligence. Referring to the 'unorganised' machine as a 'child-machine,' he wrote that he had 'succeeded in teaching it a few things, but the teaching method was too unorthodox for the experiment to be considered really successful' (p. 457).

The question now arises as to the importance of paper-based, as opposed to computer-based, experiments in Turing's methodology. It is known that he expected to perform experiments on digital computers as soon as they became available (see note 128). The question is, why did he eagerly await the availability of a digital computer? Was it because he thought that computer-based experiments would be more meaningful or significant? The answer is hinted at in Turing's own reports on his experiments:

I made a start on [machine organising methods that would be more nearly analogous to our 'methods of education.'] but found the work altogether too laborious at present. When some electronic machines are in actual operation I hope that they will make this more feasible.

(Turing, 1948, p. 428)

That Turing expected to use digital computers specifically to scale and accelerate his exploratory experiments, and not to elevate their status to a demonstrative role, is further emphasized in another passage where he describes '[o]ne particular kind of phenomenon' that he 'had been hoping to find in connection with the P-type machines,' which was 'the incorporation of old routines into new.' He studied this phenomenon in the context of teaching the machine arithmetic operations (e.g., multiplication as an extension of summation) and language (e.g., irregular verbs as a variation on the rules of the regular verbs). Of his partial results on both problems, he wrote respectively: 'Although I was able to obtain a fairly detailed picture of how this might happen I was not able to do experiments on a sufficient scale for such phenomena to be seen as part of a larger context;' and '[c]learly this could only be verified by very painstaking work' (p. 429).

Turing's major 1948 report on 'Intelligent Machinery' shows the systematic use of a method of successive idealizations and exploratory experiments, and does not refer to experiment as having any demonstrative role.[131] This motivates a closer look at his 1950 paper, which is the main locus of Turing's rhetoric of crucial experimentation.

6.4 Turing's 1950 paper, an exegesis

Turing's 'Computing Machinery and Intelligence' has been said to be accessible to a general readership.[132] But it has also been said to be a complex, multi-layered text (Genova, 1994), too ambiguous for interpretation (Hayes and Ford, 1995; McDermott, 2014). An intriguing puzzle about Turing's 1950 text is why he considered the original question, whether machines can think, to be 'too meaningless to deserve discussion' and proposed to replace it with a test, an experiment, and yet spent most of his paper (sections §§6, 7, almost 70% of the paper, as detailed below) discussing the question.

I will propose a way of reading Turing's 1950 text that can resolve these tensions (Section 6.4.1), and suggest a possible influence on a key aspect of its structure (Section 6.4.2).

6.4.1 *The logical structure of Turing's 1950 paper*

Turing's text, which he divided into seven sections, §1 to §7, can be read according to these three main logical steps:

- (*The proposal*, §1 to §3). A new proposal is made about how best to approach the original question, whether machines can think. One possibility is to discuss the then existing common-sense notions of machines (e.g., a steam engine) and thinking (what humans do). It is noted, however, that this would make the question paradoxical from the outset, and indeed absurd. The imitation game is introduced as an idealized scenario designed to be a sensible and appropriate substitute for what is seen as obsolete common sense. The appeal and the settings of the idealized scenario are commented on; in particular, why blind conversation makes sense as an intellectual task to empirically evaluate the cognitive abilities of digital computers (the new machines then in existence) to do something that, if done by a human being, should be called 'thinking.' The imitation game is thus presented as a vivid and picturesque image that contains an (epistemological) 'criterion for thinking.' Based on the imitation game, two variants (man-imitates-woman and machine-imitates-woman versions) of the new question are described in place of the original one. Further variations of it will be suggested as the text progresses to its next logical step.

- (*The science*, §4 to §5). The digital computer is explained in language widely accessible to readers in philosophy, mathematics, and science, if not to the general public. It is emphasized that the proposal (the first logical step) is a philosophical reflection on a science — namely, Turing's 1936 mathematical science of universal computing. This science was combined with the technology of stored programs developed in the computer building projects of the postwar years. This combination was not accidental but fine-tuned in order to make digital computers behave or perform as universal computing machines. At the end of §5, 'Universality of Digital Computers,' it should be clear that the proposal does not suggest some arbitrary idealized scenario, but one that is informed and constrained by the science and technology of digital computers. (The integration of common sense and cutting-edge science is a hallmark of thought experiments in the scientific tradition.)

- (*The discussion*, §6 to §7). On the original question of whether machines can think, a negative (§6) and a positive (§7) argumentation is presented using the scientifically informed proposal. First, Turing's beliefs and views are explained. The scientific status of the question is considered open. Turing's own conviction is that the answer to the question is positive, but he would rather avoid saying so directly, for the very reason that he has outlined the proposal in the first place—namely, to provide a basis for the discussion not to be meaningless. The discussion itself unfolds by addressing each of a series of nine objections to the possibility of an intelligent machine, systematically referring back to the imitation game. This is the negative argument. Then a research project is presented for the development of 'learning machines' that could be made to play the imitation game well. These, once provided with the necessary storage capacity and a suitable program, would be able to learn by themselves. By analogy with the education of a human child, the proposed approach is to find a 'child program' that would have little structure at first and would grow in complexity with the machine's experience, so that the machine could eventually exhibit its own intelligence in the imitation game. This is the positive argument. It is worth noting that the whole discussion is developed on the basis of the imitation game.

It is worth noting that the focus of Turing's argument, as indicated by the development of the text, is *not* on the first step, the proposal, but on the third step, the discussion of the original question whether machines can think. (Turing's argument reaches its climax in section §6 and its full development in section §7.) Without the proposal (first step), however, the discussion would be based not on science but on culture, namely, the commonsense notions of 'machine' and 'thinking' at the time.[133] Given the intended conceptual change in the meaning of these words based on a new science, this would indeed be absurd. Furthermore, without an explanation of the new science (second step), the proposal could be understood as nothing more than an imaginary exercise in fiction or fantasy. But having established such basic premises, the discussion could finally unfold. It would then have an empirical basis in the proposed (epistemological) 'criterion for "thinking."' At the same time, this criterion was embodied in a reasonable idealized scenario (the imitation game) in order to maintain an appeal to common sense. This brief analysis may facilitate the understanding that Turing's test proposal was in fact a means, not an end.[134]

We can now proceed to examine what kind of philosophical argument Turing developed at the climax of his paper, his discussion of the original question, and what is a possible source to have influenced it.

6.4.2   The Socratic dialectic method

Turing explicitly referred to Bertrand Russell's *History of Western Philosophy* (1945) in his discussion of the 'theological' objection (1950, p. 443). Russell's book had appeared only five years earlier, and is one of the few works in the bibliography of Turing's 1950 paper, the only work from philosophy.

Russell (1872–1970) wrote a chapter on Socrates, from which I will generously quote, in which he introduced the Socratic method of dialectic:

> Dialectic, that is to say, the method of seeking knowledge by question and answer, was not invented by Socrates ldots But there is every reason to suppose that Socrates practised and developed the method ldots Certainly, if he practised dialectic in the way described in the [*Apology*], the hostility to him is easily explained: all the humbugs in Athens would combine against him.
>
> (Russell, 1945, p. 92)

Parallel to Russell's introduction of 'the method of seeking knowledge by question and answer,' Turing would justify his choice of the intellectual task to be addressed in the imitation game: 'The question and answer method seems to be suitable for introducing almost any one of the fields of human endeavour that we wish to include' (1950, p. 435). And if there was hostility to Socrates in Athens, there was certainly hostility to Turing in postwar

Britain, leading him to title one of his BBC radio broadcasts 'Intelligent Machinery, a Heretical Theory' (1951a).

Russell also noted that the dialectical method was the method used by Galileo in his dialogues to argue his theories and overcome prejudices. Russell went on to reflect on the limitations of the method as paradigmatically demonstrated by Galileo. Note the parallel, as Turing would celebrate Galileo at the end of his rebuttal of the theological objection,[135] which is precisely where he cites Russell's *History*, and later again in his 1951 BBC radio lecture 'Intelligent Machinery, a Heretical Theory.'[136] Specifically, Russell wrote:

> The dialectic method is suitable for some questions, and unsuitable for others ... Some matters are obviously unsuitable for treatment in this way — empirical science, for example. It is true that Galileo used dialogues to advocate his theories, but that was only in order to overcome prejudice — the positive grounds for his discoveries could not be inserted in a dialogue without great artificiality. Socrates, in Plato's works, always pretends that he is only eliciting knowledge already possessed by the man he is questioning; on this ground, he compares himself to a midwife.
>
> (Russell, 1945, pp. 92–93)

Russell refers to the image of Socrates as a midwife to make the point that philosophical discussion can produce conceptual clarity, but it cannot produce new knowledge about the natural world. He went on to explicitly state that the Socratic method of dialectic does not apply to empirical problems, such as 'the spread of disease by bacteria.' He advised against using philosophical discussion as if it could provide any positive grounds for discovery about the natural world.

The parallel with Turing's argumentative approach as shown in the third logical step of his 1950 text ('the discussion') can be revealing. In the positive part of his discussion (§7 of his text), Turing would write: 'The reader will have anticipated that I have no very convincing arguments of a positive nature to support my views.' He added: 'If I had I should not have taken such pains to point out the fallacies in contrary views.' Whether under the influence of Russell's *History* or not, it can be seen that Turing revived the Socratic dialectical approach, while respecting the limits suggested by Russell and allegedly shown by Galileo regarding any positive claims to his empirical question of whether machines can think.

Russell concluded by emphasizing that the proper use of the Socratic method is for questions about the meaning and use of words:

> The matters that are suitable for treatment by the Socratic method are those as to which we have already enough knowledge to come to a right

conclusion, but have failed, through confusion of thought or lack of analysis, to make the best logical use of what we know. A question such as 'what is justice?' is eminently suited for discussion in a Platonic dialogue. We all freely use the words 'just' and 'unjust,' and, by examining the ways in which we use them, we can arrive inductively at the definition that will best suit with usage. All that is needed is knowledge of how the words in question are used. But when our inquiry is concluded, we have made only a linguistic discovery, not a discovery in ethics.

(Russell, 1945, p. 93)

The meaning and common use of the words 'machine' and 'thinking' or 'intelligence' were just the central topic of Turing's 1950 paper. Nonetheless, again in parallel with Russell's point that no positive discovery could come out of an application of the dialectical method, Turing emphasized that he did not expect to have very convincing arguments of a positive nature to support his views. He stated instead: 'The only really satisfactory support' that can be given for the belief in his prediction 'will be that provided by waiting for the end of the century and then doing the experiment described' (p. 455). Turing's use of negative dialectics 'to point out the fallacies in contrary views' seems to have compelled him to acknowledge that he was also limited by it, and thus only a practical experiment could provide satisfactory support for his views.

For an example of what Turing considered to be a logical fallacy of the contrary views, see the fourth objection, the 'Argument from Consciousness' (1950, p. 446). Turing casts of Geoffrey Jefferson's demand for machine sentience as a condition for machine intelligence as an instance of solipsism with no basis in common sense. Here Turing fits the Socratic image of a midwife supposedly giving birth to truth. In fact, Russell's point about the Socratic method as capable of producing truth after correcting 'logical errors' might have been compelling to Turing:'

We can, however, apply the method profitably to a somewhat larger class of cases. Wherever what is being debated is logical rather than factual, discussion is a good method of eliciting truth. Suppose some one maintains, for example, that democracy is good, but persons holding certain opinions should not be allowed to vote, we may convict him of inconsistency, and prove to him that at least one of his two assertions must be more or less erroneous. Logical errors are, I think, of greater practical importance than many people believe; they enable their perpetrators to hold the comfortable opinion on every subject in turn. Any logically coherent body of doctrine is sure to be in part painful and contrary to current prejudices. The dialectic method ... tends to promote logical consistency, and is in this way useful. But it is quite unavailing when the object is to discover new facts.

(Russell, 1945, p. 93)

In summary, Turing separated his use of negative dialectic to discuss the original question as a logical question (§6) from his proposal of a research project to address it as an empirical question (§7). He referred to the imitation game as a criterion by which he could expose such logical fallacies, since it placed machines and humans on the same level, side by side.

6.4.3 Archival source

Following the clue of Turing's explicit citation of Russell's *History* as one of the few works in his bibliography, we have seen the analytical similarity between Turing's philosophical approach in his 1950 paper and the recommended method for philosophical discussion in Russell's *History*. Moreover, there is an archival finding which can further support my hypothesis that Russell's *History* may have been a source for Turing's philosophical approach.

A letter to Turing from a close friend of Russell, Rupert Crawshay-Williams (1908–1977), who would publish his *Russell Remembered* in 1970, describes Russell's reception of Turing's 1950 paper:

> Dear Turing,
>   I meant ages ago - to thank you for sending me the offprint of your *Mind* article. ... I am most pleased to have it, as I enjoyed it very much when it first came out. And, you may be amused to hear, so did Bertrand Russell who was here at the time. We read it and discussed it together. We liked not only (of course) the general approach (the assumptions underlying your argument) but also the particular method and the examples. How did you discover about the Encyclopedia Britannica? ...
>   I hear you've been made F.R.S. Many congratulations. But as I don't know whether it's official yet I won't stick it in the envelope.
>   Yours sincerely,
>   Rupert Crawshay-Williams[137]

Crawshay-Williams' reference to the possibility of Turing being amused to hear of Russell's appreciation of his paper and the 'of course' note suggests a strong connection with Turing's 'general approach,' 'the assumptions underlying his argument.' Their shared appreciation of 'the *particular* method and the examples' used by Turing (emphasis added) is also significant. Beyond Turing's acquaintance with Russell's philosophy as part of the

Cambridge milieu, the letter may also indicate some prior context regarding Turing's paper.

It may also be suggestive that the same letter that informs Turing of Russell's reception of his paper is the one that congratulates Turing on being made a Fellow of the Royal Society of London for Improving Natural Knowledge. Russell was an FRS, and the Royal Society was an institution known for its support of experiment as the primary source of knowledge about the natural world.[138] Turing probably wrote his paper in early January 1950, and in March 1951 he would be elected an FRS with the support of two sponsors, Russell and Max Newman.[139] Far from suggesting that satisfying Russell and others was a primary goal for Turing, I suggest that Turing cared about being understood, and that he knew how to adapt his language to address specific readers, and he had Russell as an intended interlocutor.

Turing was considered an avowed non-scholar in the sense that he avoided relying on the work of others, preferring to figure things out and solve problems on his own (Newman, 1955). However, when it came to being read and understood, he certainly showed interest in the reception of his work. For example, while in Princeton, in the United States, he carefully instructed his mother to distribute offprints of his 'Computable Numbers' (1936) in England.[140] One of the six people on his list, and the one who deserved special instructions, was Russell. Believing that Russell was, in Turing's words, 'inclined to be ashamed of his peerage,' Turing thought 'the situation calls for tact.' He added: 'I suggest that the correct address for an earl be used on envelope, but that you mark the reprint itself 'Bertrand Russell' on the top right hand corner of the cover.'

Galilean resonances can be found both in Turing's method of successive idealizations and exploratory experiments, and in his rhetoric of experimentation.

6.5   Galileo's construction of idealized fall in a void

According to the story of the Leaning Tower of Pisa, one of the most famous anecdotes in the history of science, sometime around the year 1590 Galileo would have climbed to the top of the tower and dropped two unequal weights. He did this, so the story goes, to disprove Aristotle's law of fall, which claimed that the speed at which bodies fell was proportional to their weight. By showing that the objects reached the ground simultaneously, Galileo would have demonstrated to the professors and students gathered around the tower that Aristotle was wrong. This legendary story will provide us with a case study in the history and historiography of science whose parallel to the problem of Turing's rhetoric of experimentation may be significant.

The story is never mentioned in Galileo's own writings. It was reported 12 years after his death by one of his closest students and collaborators, Vincenzo Viviani (1622–1703), as part of a biography of Galileo written in 1654 and first published posthumously in 1717.[141] The Leaning Tower demonstration has often been considered a turning-point in the history of science, and many authors who believe that Galileo's science was mainly empirical have produced it as a classic example of the superiority of empirical science over *a priori* science. For centuries, Galileo was largely considered 'the first true empiricist' (cf. Segre, 1989b, p. 207), which may explain Russell's account described above. Writing in the early 1940s, Russell could hardly be aware of the studies of Lane Cooper (1935) and Alexandre Koyré (1937, 1939) in the 1930s, whose shockwaves would be felt in Galileo scholarship for decades to come (Koyré, 1953; Settle, 1961; Shea, 1972; MacLachlan, 1973; Drake, 1973; Naylor, 1974, 1976; Drake, 1978; Adler and Coulter, 1978; Franklin, 1979; Segre, 1980, 1989a; Palmieri, 2005a,b).

In summary, despite the beliefs of Stillman Drake (1978),[142] it is unlikely that Galileo performed the legendary tower experiment as described by Viviani (cf. Cooper, 1935; Koyré, 1937; Segre, 1989a). And if he did, he could hardly have obtained the claimed results under the conditions available to him at the time (cf. Naylor, 1974; Adler and Coulter, 1978; Segre, 1980). But there is a caveat. Galileo seems to have done various exploratory experiments. He seems to have used what he learned from such experiments to infer what should be expected in scenarios and conditions whose ideal experimental conditions were beyond his reach. He combined reason with experiment in his theory-building process, and used conceptual experiment in his dialogical arguments. A corollary of this, as suggested by Segre in the Galilean case (1980), is that 'we must notice that not all experiments are meant to be performed' (p. 228). (I will suggest in Section 6.6 that Segre's point can also apply to the Turing case, despite Turing's rhetoric about performing a crucial experiment.)

In the following, I will briefly discuss the findings of Galilean studies in relation to two questions. The first question is: (Section 6.5.1) How did Galileo establish his law of free fall? Did he rely on crucial experiments, as Russell's account of him as an empiricist might suggest? The second shifts the problem from the history of science to its historiography: (Section 6.5.2) Why was a rhetoric of crucial experiments important to Galileo and to his student and first biographer, Vincenzo Viviani? The answers provided by Galilean studies can shed light on the problem of Turing's rhetoric of experimentation.

6.5.1   *Galileo and the law of free fall*

According to Paolo Palmieri (2005b), Galileo's journey to formulate 'the most beautiful thought experiment in the history of science' in his *Two New Sciences* (1638) began some five decades earlier in the drafting of his unpublished *De Motu* (1590). Palmieri reconstructs Galileo's investigations from his early Archimedean arguments about floating bodies and finds him discovering paradoxical phenomena in Aristotle's arguments against the possibility of the void. Galileo would have started from the analogy that the reason why objects of the same kind, though different in volume, fall at the same speed is the same as the reason why both a chip of wood and a large wooden beam float. The reason why the beam behaves the same as that of the chip is that both must lift a quantity of water equal to their volumes as they fall (1909, vol. I, pp. 263-264). Following this line of reasoning, Galileo would eventually come up against Aristotle's claim that if motion occurred in the void, then both heavy and light bodies would move at the same speed, since the resistance of the void to their motions would be zero, which is unnatural (*inconvenience*) (Galilei, 1909, vol. I, p. 401). Aristotle's reasoning, according to Galileo, is that faster bodies cut through the medium more strongly, but since the void exerts no resistance, all motions must occur at the same speed. However, while Aristotle would have seen this as a proof of the impossibility of motion in the void, Galileo recasts it as a reason to suspend the common sense that heavy and light bodies move at speeds proportional to their weights (Galilei, 1590, p. 34).

Fast-forwarding five decades to *The Two New Sciences*, Galileo formulates his famous thought experiment through the character of Salviati:

> SALVIATI: But without experiences, by a short and conclusive demonstration, we can prove clearly that it is not true that a heavier moveable is moved more swiftly than another, less heavy, these being of the same material, and in a word, those of which Aristotle speaks. Tell me, Simplicio, whether you assume that for every heavy falling body there is a speed determined by nature such that this cannot be increased or diminished except by using force or opposing some impediment to it ...
>
> [SIMPLICIO acquiesces]
>
> Then if we had two moveables whose natural speeds were unequal, it is evident that were we to connect the slower to the faster, the latter would be partly retarded by the slower, and this would be partly speeded up by the faster ...
>
> [SIMPLICIO agrees again]
>
> But if this is so, and if it is also true that a large stone is moved with eight degrees of speed, for example, and a smaller one with four [degrees], then

joining both together, their composite will be moved with a speed less than eight degrees. But the two stones joined together make a larger stone than the first one which was moved with eight degrees of speed; therefore this greater stone is moved less swiftly than the lesser one. But this is contrary to your assumption. So you see how, from the supposition that the heavier body is moved more swiftly than the less heavy, I conclude that the heavier move less swiftly.

(Galilei, 1638, p. 66–67)

Even if the premises are true, Michael Stuart (2020) argues, following Tamar Gendler (1998), Galileo's conclusion need not follow, 'as it is open to the Aristotelian to distinguish between the weights of *united* and *unified* entities or to deny that composite objects have determinable weights' (Stuart, 2020; Gendler, 1998, p. 973; p. 405, their emphasis). However, although Galileo's use of imagination can be seen as epistemically unjustified, Stuart continues, it led to a change in the natural interpretation of the phenomenon, which allowed for its retrospective justification. An epistemic use of imagination may be considered invalid at one time, but depending on its future consequences, Stuart argues, its status may be reversed at another time (p. 974). Overall, there are uses of imagination 'that are needed to break today's constraints in order to make progress tomorrow.'

In *Postils to Rocco* (1909, vol. VII, p. 731), Galileo offers an *ex post facto* reconstruction of his discovery process, writing 'I formed an axiom such that nobody could ever object to...' (The axiom corresponds to a slightly more general version of Salviati's first assumption quoted above, which Simplicio accepts.) Galileo continued to outline how he deduced the above *reductio ad absurdum*, opening the way for the law of free fall and, more importantly, for the possibility of motion in a void. According to Palmieri (2005b, p. 231), Galileo suggested that it was reason, not experience, that convinced him of the law of free fall for a certain class of phenomena (when bodies of different weights but the same material). Nearly five decades later, then in possession of the law of accelerated fall, and of a complex theory of fluid resistance, Palmieri argues (2005b, pp. 224–225), Galileo would have 'recast his original thought experiment into the punchy presentation that was eventually published in *Two New Sciences*' (1638). As noted by Naylor (1976), Galileo made quite distinctive uses of (real) *experiment* and (didactic) *demonstration* (my emphasis).

Now, note that Galileo's reduction suggests an anomaly in Aristotle's theory of motion, but it cannot, in Russell's terms, provide a positive ground for Galileo's own (existential) hypothesis, namely, that there can be motion in a void. Strictly speaking, the empirical study of this hypothesis within a given location requires the removal of the medium from that location. In the specific case of the nearly 200-foot Leaning Tower of Pisa, this would require a chamber the size of the tower and the removal of all air from it. Thus, the empirical evaluation of Galileo's hypothesis at that time would involve begging the question or assuming the conclusion, since creating the conditions that would make the crucial experiment feasible would require pursuing the hypothesis in question through long-term research and development.[143] Historically, by the time a crucial experiment was possible, the related Galilean science was so far advanced that there was little point in doing the experiment except to honor Galileo or for entertainment. It was not until the space programs of the second half of the twentieth century, which were themselves based on Galilean science, that the ideal conditions became available to carry out the legendary tower experiment.[144]

I do not intend to suggest that experiment did not play an important role in Galilean science, for it did (cf. Naylor, 1974, 1976), but only that it did not play a *demonstrative* role (Segre, 1980, my emphasis). According to Palmieri (2005a), elaborating on William Shea (1972), the methodology of Galileo's intellectual revolution consisted basically in the mathematical study of classes of phenomena under certain idealized conditions. Galileo's investigations relied heavily on thought-guided experiments, some of which also involved what Palmieri calls 'participation, namely, material or bodily activities' (2018). The intriguing relationship between different experimental settings and scenarios used by Galileo had been observed earlier by Ernst Mach (1897). Mach found that continuous variation between various conditions and scenarios is the fundamental method of both thought and physical experiment. This helps to explain the close relationship between some of Galileo's experiments. For example, Koyré (1953) notes, 'It is well known with what extreme ingenuity, being unable to perform direct measurements, Galileo substitutes for the free fall the motion on an inclined plane on one hand, and that of the pendulum on the other' (p. 224).

So much for the first question on how Galileo would have established his law of free fall, we can now move on to the second. If Galileo did not rely on crucial experiments, why was a rhetoric of crucial experimentation important to Galileo, or at least to his student and biographer, Vincenzo Viviani? The legendary tower experiment provides us with a particularly interesting case study of the rhetoric of crucial experiments in the history and historiography of science.

6.5.2   The social and cultural dimension of the legend of Galileo's tower experiment

Early on in his *De Motu* writings (1590), Galileo reports on tower experiments.[145] However, as Segre (1989a) notes, in these specific reports Galileo repeatedly and explicitly stated that bodies of different weights fall at different speeds. For example, he reports dropping two different bodies, one of lead and one of wood, from the top of a high tower: 'the lead moves far out in front. This is something I have often tested' (p. 107). And yet the biographer Vincenzo Viviani, who studied with the late Galileo, would describe Galileo's feats in Pisa according to Galileo's mature views:

> And then, to the dismay of all the philosophers, very many conclusions of Aristotle were by him [Galileo] proved false through experiments and solid demonstrations and discourses, conclusions which up to then had been held for absolutely clear and indubitable; as, among others, that the velocity of moving bodies of the same material, of unequal weight, moving through the same medium, did not mutually preserve the proportion of their weight as taught by Aristotle, but all moved at the same speed … demonstrating this with repeated experiments from the height of the Campanile of Pisa in the presence of the other teachers and philosophers, and the whole assembly of students.
>
> (Galilei, 1909, vol. XIX, p. 606)[146]

Faced with Viviani's embellishment of Galileo's exploits, Segre contributed an important insight by asking the historiographical question, 'Why did Viviani think it important to report such an experiment?' He looked at Viviani's literary context and found that in Viviani's milieu a biography had to follow certain standards. The paradigmatic example was the *Vite*, the famous collection of biographies in art history written by the Mannerist painter and architect, Giorgio Vasari (1511–1574). Segre found concrete examples of Vasari's form and style in Viviani's biography of Galileo, including the embellishment of the portrayed artist's image with quasi-true anecdotes. Moreover, Viviani's intended audience, after Galileo himself in his lifetime, included the general educated public (the nobility, the learned clergy, and the academic community). Segre highlighted an example of the expectations of such an audience in the correspondence between Galileo's followers, Bonaventura Cavalieri (1598–1647) and Evangelista Torricelli (1608–1647). Cavalieri wrote to advise Torricelli on the occasion of the latter's admission to the *Accademia della Crusca*: 'I hear that they expect physical rather than mathematical things … It is advisable to meet their expectation, and more than that, the universal expectation that has little esteem for mathematics, unless it sees some applications' (cf. Segre, 1989a, p. 447). Segre further noted that Viviani revised the first version of his Galileo biography to emphasize the tower anecdote over the abstract thought experiment that sought to refute Aristotle, probably in an attempt to be more convincing to such an audience. In sum, the anecdote of the leaning tower experiment was an important device for providing the general educated public with a physical, tangible story and application of the otherwise overly abstract Galilean science. Although the anecdote may have been invented by Viviani and not by Galileo himself (*ibid.*, pp. 441, 444), we can see in Galileo's works themselves, for example in the *Dialogo* and the *Discorsi*, as noted by Naylor (1976), Galileo's 'ambition to extend the results of his original experiments to important practical problems' (p. 400).

Centuries later, the 'Apollo 15 Hammer-Feather Drop' was a live anecdotal demonstration of Galileo's legendary tower experiment by mission commander David Scott for the television cameras at the end of the last Apollo 15 moonwalk on August 2, 1971. Far from the Earth's atmosphere, essentially in a vacuum, the astronaut simultaneously released a heavy object (an aluminum geological hammer) and a light object (a falcon feather) from approximately the same height (about 1.6 m), which fell to the ground at the same rate to the naked eye. The performer, who attributed their successful mission in part to 'a rather significant discovery about falling objects in gravity fields' made long ago by 'a gentleman named Galileo,' celebrated: 'How about that! Mr Galileo was correct in his findings.'[147]

Undoubtedly, there is a remarkable historical connection between the Apollo 15 hammer-feather drop and Galileo's public demonstrations and legends, especially the story of the Leaning Tower of Pisa experiment. Robert Crease (2003) pointed out that Galileo's experiments 'slowly transformed from genuine scientific inquiries into public displays,' greatly influencing a next generation of scientists. These included Robert Boyle (1627–1692) and Willem's Gravesande (1688–1742), who built air pumps and special chambers to study vertical fall in evacuated environments. Authorities such as King George III, for example, once witnessed a demonstration of a feather and a one-guinea coin falling together inside an evacuated tube. 'The popularity of such demonstrations,' Crease notes, 'continues to this day' and is included in many hands-on science exhibits. For him, even if there was no original (leaning tower) experiment, Galileo inspired a whole genre of experiments and demonstrations, and 'we might as well refer to these as the offspring of Galileo's experiment at the Leaning Tower of Pisa.' Boyle would indeed become a master at using instruments to present to the public indisputable facts produced in crucial experiments (Shapin, 1984). And he certainly helped promote Galileo's hypothesis that there could be motion in a void. As a teenager, Boyle visited Florence with his French tutor shortly before Galileo's death and was impressed by 'the new paradoxes of the great star-gazer Galileo' (Fulton, 1960, p. 119). If Boyle was recruited and pushed forward a rhetoric of crucial experimentation, the Galilean legends and propaganda seem to have contributed to it.[148]

Segre (1980) emphasized the importance of distinguishing between Galileo's method and his methodology, in other words, in practice Galileo's method 'was not a direct outcome of his methodology and vice-versa (p. 249). While Galileo, the scientist, relied largely on a method of successive idealizations and exploratory experiments, and resorted to thought experiments for didactic demonstration, Galileo, the methodologist, promoted a view of himself as an empiricist at strategic moments, and resorted to a rhetoric of decisive experiments and propaganda. This chapter proposes to apply a similar distinction to Turing.

6.6 Turing's use of thought experiment and propaganda

In light of the historical case of Galileo, we can now revisit Turing's construction of machine intelligence to emphasize his use of thought experiment (Section 6.6.1) and propaganda (Section 6.6.2).

6.6.1 *Turing's use of thought experiment*

By 1948, as we have seen, Turing had already developed his concepts of machine intelligence and imitation testing. In the absence of actual digital computing equipment to scale his initial experiments, he explored machine intelligence by simulating on paper the learning process of successive models of his 'unorganized' machine.

Turing also explored human perception of machine intelligence through the abstraction of a chess-playing paper machine. 'Playing [chess] against such a machine,' Turing wrote, 'gives a *definite* feeling that one is pitting one's wits against something alive' (p. 412, my emphasis). Writing in 1948, it seems that Turing himself needed no further evidence to believe in the possibility that average interrogators in the future could be fooled by a sophisticated machine in a third of the experimental sessions after five minutes of questioning. In 1950, he would outline his 'beliefs' and conclude the point stating that '[c]onjectures are of great importance since they suggest useful lines of research' (p. 442). If Turing was himself already convinced, then there may be no reason to understand his 1950 crucial experiment as something other than a rhetorical strategy. Arguably, it was more a concession to meet the standards of his interlocutors than his own, which resonates with Galileo's moves from his real, relatively messy exploratory experiments to his sanitized, didactic demonstration.[149] Moving from chess to conversation, Turing would make the step from his exploratory experiments in 1948 to his thought-led imitation tests in 1950.

Chapters 4 and 5 historicized how this came about in the context of a controversy about minds and machines, and this is consistent with the parallel suggested here with Galileo's cosmological polemic. The idealizations that Turing introduced in his 1950 imitation tests can be regarded, as Karl Popper proposed in his *Logic of Scientific Discovery* (1959), as 'concessions to the opponent, or at least acceptable to the opponent' (p. 466). For example, the change from chess to conversation was probably largely a response to criticism he had received from Michael Polanyi and Geoffrey Jefferson, both Fellows of the Royal Society, sacrificing his caveat that 'many people' thought that 'a very abstract activity, like the playing of chess, would be best' (1950, p. 460).[150] He also introduced a third player, representing the imitated type, to help the human judge to identify the imitating machine. He seems to have done this, somewhat satirically, to address another point raised in Jefferson's criticism of Grey Walter's mechanical tortoises. Overall, it can be seen that Turing used his imitation tests both to criticize opposing views and to heuristically present his own views. He varied his imitation tests by varying their experimental conditions and settings, which, as Mach describes (1897), extends 'the scope of ideas (expectations) tied to them.'

When it comes to experimentation, Turing's method was to design and conduct initial experiments to explore the phenomena under study and to refine his hypotheses and analogical models. Thus experiment played an important role in Turing's actual construction of machine intelligence and imitation tests, but apparently not a demonstrative role. Turing's overarching method for establishing rigorous results was the axiomatic method, not the experimental method. In his obituary of Turing, Max Newman emphasizes Turing's reliance on mathematical proof:

> The central problem with which he started, and to which he constantly returned, is the extent and the limitations of mechanistic explanations of nature. His way of tackling the problem was not by philosophical discussion of general principles, but by mathematical proof of certain limited results ...
>
> (Newman, 1955, p. 256)

Another testimony was given by R. K. (Richard) Livesley, a graduate of the Mathematics Department at Manchester University who met Turing regularly in the Computing Machine Laboratory from 1951 to 1954 (Lavington, 2019, pp. 38–39). According to Livesley, 'Turing was fond of saying "An ounce of mathematics is worth a ton of computing."'[151] Compare Galileo in note 149, following his mathematical proof of the properties of projectile launching.

Further examining Turing's writing in 1948, it was 'the actual production of the machines'—not a crucial experiment—that could outweigh the reactions to the possibility of machine intelligence, some of which he expected to be 'purely emotional:'

> The objections (a) and (b) [which in 1950 he called the 'Heads in the Sand' and the 'Theological' objections, respectively], being purely emotional, do not really need to be refuted. If one feels it necessary to refute

them there is little to be said that could hope to prevail, though the actual production of the machines would probably have some effect.

(Turing, 1948, p. 411)

This can be read as suggesting that, for Turing, the best demonstration of machine intelligence would come with its scientific and technological development.[152] Turing's postwar career shows that he was committed to the project of building an intelligent machine.[153]

Segre (1997) notes (p. 498) that Galileo 'was, no doubt, a master of rhetoric,' but his arguments were 'not meant only to persuade.' Here, another parallel can be helpful, for Turing's belief in 'the actual production of machines' for persuasion recalls figures such as the French inventor and engineer Jacques de Vaucanson (1709–1782), famous for his artificial duck. The designers of early modern automata in the seventeenth and the eighteenth centuries were criticized for being driven by futile, purely entertaining motives. Historians of technology have disputed this. For example, David Fryer and John Marshall (1979) criticized the 'claim that the primary objective of Vaucanson's work was 'to astonish and amuse the public' (pp. 267–268). 'Vaucanson,' they argued, 'was an entertainer, but he was also deeply committed to the development of an explanatory psychology.' This analysis of Vaucanson's motives belongs to a class of studies presented earlier by Silvio Bedini (1964) and Derek de Solla Price (1964), which showed that early modern automata were neither 'trivial toys' nor 'immediately useful inventions.' Rather, they were simulacra or models 'whose very existence offered tangible proof, more impressive than any theory, that the natural universe of physics and biology was susceptible to mechanistic explication' (Price, 1964, p. 9).[154] Turing's project to demonstrate an intelligent machine can be understood in this tradition.

But how could Turing best contribute to such a project at a time when, as he himself explained, machines lacked sufficient storage capacity and computing power, and the science of machine learning that he proposed was still in its infancy? There is a secondary source known to Turing scholars which suggests an answer to this question.

6.6.2   *Turing's use of propaganda*

At this point, it is important to recall Gandy's anecdote, which further supports the approach taken in this chapter to distinguish the logical structure of Turing's 1950 paper from its rhetoric.[155] It also supports an understanding of Turing's test more as a philosophical argument and less as a crucial experiment to decide about the possibility of machine intelligence. Turing dressed his thought experiment in the imitation game, inspired by popular

material culture (Chapter 5, Section 5.5.2), and in the few years that followed he went on public radio three times to argue for his hypothesis that
machines can think (Turing, 1951a,b; Turing et al., 1952). So the record
suggests that Turing was interested in public engagement and propaganda,
as was the case with Galileo's biographer Viviani, who largely followed
Galileo himself (Segre, 1988).

Turing's test proposal seems to have influenced a next generation of
scientists, notably John McCarthy (1927–2011), who wrote (1956): 'The
problem of giving a precise definition to the concept of "thinking" and of
deciding whether or not a given machine is capable of thinking has aroused
a great deal of heated discussion.' 'One interesting definition,' he continued,
'has been proposed by A. M. Turing.' McCarthy also attributed to Turing
'[t]he first scientific discussion of human level machine intelligence' (2007),
referring to Turing's lecture to the London Mathematical Society in (1947).
'The notion was amplified as a goal in [Turing's 1950 paper],' McCarthy
added. Thus, McCarthy referred to the Turing test both as a definition of the
concept of thinking (early in his career) and as a research goal for human-
level machine intelligence (later in his career).

Marvin Minsky (1927–2016), a pioneer with McCarthy in the field of
artificial intelligence, commented to the media on the results of the practical implementation of the Turing test announced in 2014 (Section 6.1):
'Nothing is learned from poorly designed "experiments." Ask the program
if you can push a car with a string. And, if not, then, why not?'[156] A year
earlier, Minsky had said in another interview: 'The Turing test is a joke, sort
of, about saying "a machine would be intelligent if it does things that an
observer would say must be being done by a human;" so it was suggested
by Alan Turing as one way to evaluate a machine but he had never intended
it as the way to decide whether a machine was really intelligent ....'[157]

It is worth noting that McCarthy and Minsky, leading scientists of the
next generation after Turing, had no problem in distinguishing between the
substance of Turing's 1950 paper and its rhetoric of a crucial experiment.

Less than two decades after Turing's 1950 paper, his imitation test was
on its way to becoming an icon of popular culture. It was featured in Arthur
Clarke's influential novel, *2001: A Space Odyssey*:

> The sixth member of the crew cared for none of these things, for it
> was not human. It was the highly advanced HAL 9000 computer, the
> brain and nervous system of the ship ... Whether HAL [Heuristically
> programmed ALgorithmic computer] could actually think was a ques-
> tion which had been settled by the British mathematician Alan Turing
> back in the 1940s. Turing had pointed out that, if one could carry out
> a prolonged conversation with a machine—whether by typewriter or
> microphone was immaterial—without being able to distinguish between

its replies and those that a man might give, then the machine was thinking, by any sensible definition of the word. HAL could pass the Turing
test with ease.

(Clarke, 1968, p. 97)

Hodges (1983, p. 533) asked whether Clarke might have chosen the year
2001 to fulfill Turing's prediction quoted here at the beginning (Section 6.1).

Here I have suggested a parallel between Turing's and Galileo's methods, but in fiction and popular culture the products of their science and
propaganda meet.

6.7   Conclusion

Turing articulated a rhetoric of a crucial experiment to determine the existence of an intelligent machine (Section 6.1). And clearly, experiment played
an important role in Turing's construction of machine intelligence (Section 6.3). However, it can be seen that his rhetoric of crucial experimentation appears as a specific move in his 1950 paper, which is followed in later
primary sources by a milder attitude towards the confirmatory power of
what he called in 1952 his 'imitation tests.' Earlier, in his major 1948 report
on 'Intelligent Machinery,' Turing refers to the demonstrative power of 'the
actual production of machines' rather than that of a controlled experiment,
somewhat alluding to the persuasive power of scientific and technological
progress itself.

These tensions in Turing's primary sources motivate a distinction
between the logical structure of the Turing test argument and Turing's
rhetoric of experimentation. An exegesis of Turing's seminal 1950 paper
has been proposed (Section 6.4), identifying in it Turing's philosophical
discussion of the question 'can machines think?' as its main logical step. I
have argued, both analytically and historically, that Turing's discussion may
have been influenced by Bertrand Russell's account of the Socratic dialectical method in his *A History of Western Philosophy*. Further, I have argued
that Russell, and perhaps Turing's context more generally, including other
interlocutors and the Royal Society as an institution supporting controlled
experimentation as the primary source of knowledge about the natural
world, can partly explain Turing's rhetoric of experimentation. The latter
may have been a concession to meet the standards of his interlocutors more
than his own.

The frame of analysis has been broadened by looking at the maturity
of Galilean studies. I have drawn a parallel with Galileo's construction of
idealized fall in a void and have briefly reviewed the historiographical controversies about the role of experiment in Galilean science (Section 6.5).
We have seen that Galileo could hardly have obtained the results claimed

by the story of the Leaning Tower experiment. To perform such an experiment under the ideal conditions necessary to validate Galileo's existential hypothesis—namely, that there can be motion in a void—was beyond the possibilities at that time. The empirical evaluation of Galileo's hypothesis implied begging the question or assuming its conclusion, since making the crucial experiment feasible required pursuing the hypothesis itself through long-term research and development. Both in the secondary literature and in Turing's own statements, there are indications of the perception that Turing's imitation tests are circular. As has been argued in the Galileo literature, not all experiments are meant to be performed. This seems to be true for the Turing test.

Finally, I have emphasized in Turing's construction of machine intelligence his use of thought experiment and propaganda (Section 6.6). This resonates with important findings in Galileo scholarship, particularly the proposed distinction between the method Galileo used and his alleged methodology. Turing, comparably with Galileo, I suggest, recognized the need to inspire further research that could lead to convincing scientific and technological progress. Galileo's construction of idealized fall in a void, which is an important case in the history and historiography of science, shows that such tactics have been used with great success in the past to change the existing natural interpretation of phenomena. We may be on the verge of seeing history partially repeat itself with Turing's construction of machine intelligence.

7 Irony with a Point

Turing made strong statements about the future of machines in society. This chapter asks how they can be interpreted to advance our understanding of Turing's philosophy. His irony has largely been caricatured or minimized, and he is often portrayed as an irresponsible scientist, or associated with childlike manners and polite humor. While these representations of Turing have been widely disseminated, another image suggested by one of his contemporaries, that of a nonconformist, utopian, and radically progressive thinker reminiscent of the English Romantic poet Percy B. Shelley, has remained largely underexplored. Following this image, this chapter reconstructs the argument underlying what Turing called (but denied being guilty of) his 'Promethean irreverence' (1947–1951) as a utopian satire directed against chauvinists of all kinds, especially intellectuals who might sacrifice independent thought to maintain their power. These, Turing hoped, would eventually be rivaled and surpassed by intelligent machines and transformed into ordinary people, as work once considered intellectual would be transformed into non-intellectual, so-called 'mechanical' work. It is suggested that Turing genuinely believed that the possibilities of the machines he envisioned were not utopian dreams, and yet he conceived them from a utopian frame of mind, aspiring to a different society.[*]

7.1 Intelligent machinery, a heretical theory

Alan Turing (1912–1954) made two BBC radio broadcasts in 1951 that contained some of his strongest statements about the possibility of intelligent machines and their consequences for humanity. One of the broadcasts, first aired on May 15, was part of a series *Automatic Calculating Machines* that featured other British computer pioneers (Jones, 2004). The series may have limited Turing's choice of title, 'Can Digital Computers Think? (Turing, 1951a). For the other broadcast, however, Turing gave the title 'Intelligent Machinery, a Heretical Theory' (Turing, 1951b). I will quote from the climax of Turing's 1951 public lectures, starting with the former:

[*]This chapter is based on Gonçalves, B. (2023). Irony with a point: Alan Turing and his intelligent machine utopia. *Philosophy & Technology* 36, 50. doi: 10.1007/s13347-023-00650-7

If a machine can think, it might think more intelligently than we do, and then where should we be? Even if we could keep the machines in a subservient position, for instance by turning off the power at strategic moments, we should, as a species, feel greatly humbled. A similar danger and humiliation threatens us from the possibility that we might be superseded by the pig or the rat. This is a theoretical possibility which is hardly controversial, but we have lived with pigs and rats for so long without their intelligence much increasing, that we no longer trouble ourselves about this possibility. We feel that if it is to happen at all it will not be for several million years to come. But this new danger is much closer. If it comes at all it will almost certainly be within the next millennium. It is remote but not astronomically remote, and is certainly something which can give us anxiety.

> (Turing, 1951a, pp. 485–486)

Turing is concerned with our emotional response to a question we might call ontological: whether there could be some species, including machines considered as a species, that would eventually surpass the human race in intelligence. He reminds us that humans have held a dominant position among the species, but this position, he points out, is not necessarily permanent. He alludes to the pig and the rat, his carefully chosen examples, to point out, somewhat ironically, our sense of superiority over other species. He then shifts the discussion to the timescale of possible events, since the evolution of machines would not be subject to the same timescale. He warns that the threat posed to us from intelligent machines is a 'remote but not astronomically remote' possibility.

In the climax of his other 1951 BBC broadcast, Turing's focus shifts:

Let us now assume, for the sake of argument, that these [intelligent] machines are a genuine possibility, and look at the consequences of constructing them. To do so would of course meet with great opposition, unless we have advanced greatly in religious toleration from the days of Galileo. There would be great opposition from the intellectuals who were afraid of being put out of a job. It is probable though that the intellectuals would be mistaken about this. There would be plenty to do, trying to understand what the machines were trying to say, i.e. in trying to keep one's intelligence up to the standard set by the machines, for it seems probable that once the machine thinking method had started, it would not take long to outstrip our feeble powers. There would be no question of the machines dying, and they would be able to converse with each other to sharpen their wits. At some stage therefore we should have to expect the machines to take control, in the way that is mentioned in Samuel Butler's 'Erewhon'.

(Turing, 1951b, p. 475)

Here Turing is concerned with our emotional response to a question we might call ethical: whether the construction of such intelligent machines would be a sensible project. It is worth noting that Turing's focus was on a particular figure in society, the intellectual, who would be afraid of losing his or her job. He then directed the subject to the projection that once intelligent machinery is achieved, it would not take long for machines to take control, and explicitly cited Samuel Butler's Victorian novel *Erewhon* (1872) as an inspiration.

These strong statements about the future of machines in society have long been intriguing. This chapter asks how they can best be interpreted to advance our understanding of Turing's philosophy. Was Turing just joking, or was he trying to make a serious point?

7.2   Argument sketch

Biographers, historians, philosophers, scientists, and novelists have answered this question differently. I will argue that Turing's ironic and humorous attitude has led most interpreters to either caricature (as Frankenstein-like) or downplay (as polite humor) his vision of the future of intelligent machines in society, while a close and sustained examination of his arguments seems to be lacking. In particular, I will identify three images of Turing drawn from contemporaries and disseminated in the secondary literature (Section 7.3), and I will explore the first, which portrays Turing as a nonconformist, utopian, and radically progressive thinker reminiscent of Percy B. Shelley.

I will propose a historiographical and philosophical interpretation of Turing's ironic statements as satire, or irony with a point. I will suggest that Turing embraced humor and irony as a personal philosophical stance and used it as a method of self-expression (Section 7.4), making arguments through the formulation of surprising contrasts intended to unsettle the assumptions of his interlocutors. Further, it can be shown that his move in 1951 was not a thoughtless, isolated step. Instead, the same argument is consistently present in Turing primary sources every year from 1947 to 1951. I will emphasize what I call its ontological and ethical components, which appear, for example, in Turing's formulation of two objections to intelligent machines in 1948 and 1950. These objections articulate what he called—but denied being guilty of—his 'Promethean irreverence' (Section 7.5). Turing had seen models of satire in the works of Charles Dickens and Samuel Butler, whose influences can be traced in Turing's peculiar conception of an intelligent machine (Section 7.6). For a variety of reasons, I will argue that Turing's project can hardly be associated with that of Mary Shelley's character, Dr. Frankenstein (Section 7.7). I will reconstruct Turing's overarching argument, here interpreted as a utopian satire whose point is manifested in his conception of an intelligent machine (Section 7.8). Concluding remarks are given at the end, drawing a parallel between Turing's intelligent machine utopia and Percy B. Shelley's masterpiece, *Prometheus Unbound* (Section 7.9).

In addition to helping to clarify a puzzling aspect of Turing's philosophy, this chapter draws attention to an important aspect of the future of intelligent machines in society—namely, the impact of intelligent machines on the ability of intellectuals to exercise power. This is a non-obvious connection that Turing keenly foresaw. Distrustful of the attitudes of some intellectuals in positions of power, Turing hoped that his truly intelligent, ever-learning machines would expose the various forms of chauvinism he saw in their views of society and nature. Such intellectuals would eventually be surpassed by the machines he envisioned and transformed into ordinary people, as work once considered intellectual would be transformed into non-intellectual, so-called mechanical work. I study Turing's irony in its historical context and follow the internal logic of his arguments to their limit. I will suggest that he genuinely believed that his ever-learning child machines, educated by individuals (not by large corporations or nation-states) to grow their intelligence out of their own experiences, would help to distribute power in society.

7.3   Three images of Turing

Starting with Turing's contemporaries, I will move through the commentary of Turing's biographer Andrew Hodges to more recent commentators. Following this chronological approach, I will examine *three different images of Turing* that can be identified in the secondary literature in relation to Turing's use of irony (Sections 7.3.1–7.3.3). These images partly overlap and partly diverge in subtle ways: (i) a utopian, radically progressive thinker, suggesting a scientific version of the English Romantic poet Percy Bysshe Shelley (1792–1822); (ii) an infantilized, politely humorous, and muted genius; and (iii) an irresponsible scientist reminiscent of Mary Shelley's dystopian character, Dr. Frankenstein (1818). I will show that in his extraordinarily rich biography of Turing, Hodges (1983) simultaneously promotes all three images, which will testify to the fact that he offered a highly multifaceted view of Turing's character based on the testimony of Turing's contemporaries whom he interviewed. However, while two of these images have been widely disseminated, one of them, the first image of

Turing, that of a nonconformist, utopian, and radically progressive thinker reminiscent of a 'scientific' Percy B. Shelley, has remained largely underexplored since Hodges.[158] Emphasizing this image, I will follow the internal logic of Turing's own arguments to their limit.

7.3.1   First image: a nonconformist, a utopian, a sort of scientific Shelley

In a letter to Mrs. Sara Turing on December 18, 1954, a few months after her son's death in June, Geoffrey Jefferson (1886–1961), Professor of Neurosurgery at the University of Manchester during Turing's time as Reader in Mathematics there (1948–1954), offered this rich picture of Turing:

> He was so unversed in worldly ways, so childlike it seems to me, so unconventional, so non-conformist to the general pattern ... so very absentminded. His genius flared because he had never quite grown up. He was, I suppose, a sort of scientific Shelley.[159]

It is worth noting that Jefferson was born and raised in late Victorian England, just when Percy Shelley's posthumous reputation was disputed by different political forces, notably in 1886 at the foundation of the Shelley Society (Smith, 1945, p. 268), in the year Jefferson was born. Jefferson may not have understood the philosophical implications of linking Turing to the Romantic poet, for he interpreted Turing as a reductionist of mind and imagination.[160] Turing himself did not feel understood by Jefferson. According to (Hodges, 1983, p. 439), Turing 'would refer to Jefferson as an "old bumbler" because he never grasped the machine model of the mind.' It is clear, however, that Jefferson's image of Turing as a kind of scientific Shelley operates on the level of Shelley's utopianism and radical progressivism (Scrivener, 1982).

Recognized by the Lister Prize of the Royal College of Surgeons of England, Jefferson devoted his Lister Oration on June 9, 1949, in London to a critique of the possibility of thinking machines. His lecture was entitled 'The Mind of Mechanical Man' (1949a). Jefferson had joined the public discussion on mind and machine with fairly clear political concerns. Against the view he attributed to 'the physicists and mathematicians,' invaders of the field of brain-mind relations that belonged to physicians (p. 1105), he declared:

> Since no thinking man can be unaware of his fellows and of the political scene he will find that the concept of thinking like machines lends itself to certain political dogmas inimical to man's happiness. Furthermore, it

erodes religious beliefs that have been mainstays of social conduct and have brought happiness and serenity of mind to many.

(Jefferson, 1949a, p. 1107)

Jefferson's attitude can be contextualized by the Cold War climate.[161] Observing the context, we can interpret Jefferson's description of Turing without resorting to a psychologization of Turing. For someone with Jefferson's conservative views and values (Schurr, 1997), not conforming 'to the general pattern' could hardly be interpreted as something other than 'very absentminded.' This tension is largely intrinsic to Jefferson's position as a committed historical actor who perceived Turing's views as dangerous. It can be seen in Jefferson's simultaneous use of the word 'non-conformist.'[162]

Being twenty-four years older than Turing, Jefferson tried to give Turing advice, perhaps a nudge. Jefferson's opposition to Turing's views continued until their joint appearance on a BBC broadcast on January 15, 1952, which marked the end of Turing's public defense of his intelligent machine project in the media. Turing commented on the broadcast in a letter to a close friend,[163] writing that 'J. [Jefferson] certainly was rather disappointing though.' Then, in a puzzling juxtaposition, he added: 'I'm rather afraid that the following syllogism may be used by some in the future[:] Turing believes machines think. Turing lies with men. Therefore machines do not think.'

Hodges (1983), writing in the late 1970s, a few decades after Jefferson's letter of condolence, but still during the Cold War,[164] partly followed and partly departed from Jefferson in his construction of Turing's character. He explicitly acknowledged: 'Jefferson certainly found an apt description of Alan, as "a sort of scientific Shelley"' (p. 439). Here, too, Hodges seems to be largely in the mold of Jefferson, albeit with allusions to other characters:

Money, commerce, and competition had played no obvious part in the central developments in which Alan Turing was enmeshed, developments which had allowed him in many ways to remain the idealistic undergraduate. His retention of a primitive liberalism, his 'championing of the underdog' ... like his obsession with the absolutely basic, had the flavour of more Utopian thinkers than Mill. Tolstoy was a figure that he brought to one person's mind, and Claude Shannon had perceived him as like Nietzsche, 'beyond Good and Evil'.

(Hodges, 1983, p. 308)

Hodges further refined his portrayal of Turing by comparing him to the English socialist Edward Carpenter (1844–1929), one of the first socialists in Britain:[165]

But perhaps closer in spirit than either of these, and certainly closer to home, was another late nineteenth-century figure who had lurked more in the back room of political consciousness. That awkward figure Edward Carpenter, while sharing much in common with each of these European thinkers, had criticised Tolstoy for a restrictive attitude to sex and Nietzsche for overbearing arrogance. And in Carpenter, at a time when socialism was supposed to be about better organisation, lay the example of an English socialist not interested in organisation but in science, sex and simplicity—and with bringing these into mutual harmony.

(Hodges, 1983, p. 308)

Carpenter is a reference for Hodges to position Turing as more progressive than Tolstoy and less pretentious than Nietzsche. Both Turing and Carpenter lived by first principles, and Carpenter's views, Hodges notes (p. 310), played a part in 'the more innocent days' of the British Labour Party: 'His naively lucid questioning of what life was for, and of what socialism was going to make it.' At times, Hodges may give the impression that Turing was naive.

7.3.2 Second image: a genius of childlike manners, a muted figure, a gentle mocker

Jefferson's association of Turing's genius with 'childlike' ways was emphasized by Hodges (1983) in countless instances, for example in the above quote, '... allowed him in many ways to remain the idealistic undergraduate;' or, here, where he almost suggests that Turing was a rebel without a cause, 'for Alan the real point lay not in political commitments but in the resolve to question authority' (p. 72). In places, Hodges seems to agree with Jefferson's tendency to infantilize Turing, although Jefferson was acting in the role of a committed historical agent, while Hodges was playing a somewhat more detached role as Turing's post-mortem biographer. Following Jefferson, Hodges quoted from J. A. Symonds' *Shelley* (1884) to compare Turing's and Percy Shelley's attributes:

Apart from the more obvious similarities, Shelley also lived in a mess, 'chaos on chaos heaped of chemical apparatus, books, electrical machines, unfinished manuscripts, and furniture worn into holes by acids,' and Shelley's voice too was 'excruciating; it was intolerably shrill, harsh and discordant.'

(Hodges, 1983, p. 439)

However, while Jefferson considered Turing's views so dangerous that he felt compelled to respond forcefully in public, Hodges went on to write:

> Alike they were at the centre of life; alike at the margins of respectable society. But Shelley stormed out, while Alan continued to push his way through the treacly banality of middle-class Britain, his Shelley-like qualities muted by the grin-and-bear-it English sense of humour, and filtered through the prosaic conventions of institutional science.
>
> (Hodges, 1983, p. 439)

Hodges suggests that Turing's humor was polite and filtered through the conventions of his institutional environment. And yet his strong views appeared in some of Britain's most prominent media.

The depth of Turing's voice on society was not yet clear to Hodges, writing in 1983. Turing never expanded his inner circle very much, and this is a clear point that Hodges makes. However, this does not mean that Turing's radical views were silenced. Indeed, Turing envisioned machines passing one of his imitation tests around the 2050s (1952, p. 495). Meanwhile, the displacement of humans by machines has already worried neoliberal economists, who have called it the 'Turing trap' (Brynjolfsson, 2022).[166] Turing's Shelleyan qualities were not muted. However, while Hodges' foundational work itself was key to making Turing known and even a global icon, it also supported depictions of Turing that might suggest an infantilized genius, such as Derek Jacobi's portrayal in *Breaking the Code*, a 1996 BBC television movie based on the homonymous 1986 play.[167]

Jack Copeland (2012b) chose 'humour' as the first of three words, followed by 'courage' and 'isolation,' to sum up Turing: 'he had an impish, irreverent, and infectious sense of humour' (p. 1). Turing's wit fits in Copeland's biography with other qualities, 'patriotic,' 'unconventional,' and 'genius,' in the story of an unexpected protagonist of the Allied victory in the information war of World War II in the Atlantic theater. Copeland's foundational work on Turing, biographical, scientific, and philosophical, sheds further light on Turing's contributions to the Allied war effort, to early modern computing and machine intelligence, as some of Turing's contributions had long been concealed from the public record and the technical literature. But Copeland's triumphalist narrative may distract from Turing's public use of humor, which I will argue consisted of a sophisticated *defensive* tactic. I will draw attention to how Turing used his sense of humor to protect his eccentric and rebellious way of criticizing habits and customs, and social and institutional structures, particularly what Agar (2003) has called the British 'government machine.'

At a time when artificial intelligence (AI) was not as much in the public discussion as it is today, the nature of Turing's 1951 remarks on the future of machines in society was given very short shrift in Copeland's major work, *The Essential Turing*:

> Turing ends [his 1951 BBC broadcast] 'Intelligent Machinery, A Heretical Theory' with a vision of the future, now hackneyed, in which intelligent computers 'outstrip our feeble powers' and 'take control'. There is more of the same in [Turing's other BBC broadcast delivered in 1951, 'Can Digital Computers Think?']. No doubt this is comic-strip stuff.
>
> (Copeland, 2004, p. 470)

Writing in the early 2000s, when machine learning was still on the verge of becoming a dominant paradigm in AI, Copeland seems to suggest reading Turing's 1951 remarks as childish comedy.

More recently, in the wake of the recent AI resurgence, Diane Proudfoot has also commented on Turing's ironic 1951 remarks. Proudfoot focused on the alleged stupidity of Turing's interlocutors. She abstracted a notion of 'AI panic' to connect events from Butler's nineteenth century to Turing's twentieth century and contemporary events related to AI. While characterizing Turing's stance as 'gentle mockery,' Proudfoot acknowledged that it had a 'serious edge:'

> Turing (following Butler) poked fun at the fear of out-of-control AI … Turing's response to AI panic was gentle mockery. All the same, there was a serious edge to his humor. If runaway AI comes, he said, "we should, as a species, feel greatly humbled." He seemed almost to welcome the possibility of this humiliating lesson for the human race.
>
> (Proudfoot, 2015)

One question not asked is: why would Turing have welcomed a humiliating lesson for humanity? Further, was all of humanity really the target of his irony?

Overall, by emphasizing the interpretation of Turing's irony as polite humor, or by neglecting either the presence of a point in Turing's irony or the specific class or group that Turing was targeting, Turing scholars may have left the way open for more distant commentators to speculate. Turing has often been read in the same charged way that he was read in a Cold War climate, and, as will be shown below, often as a dystopian agent.

7.3.3   *Third image: an irresponsible scientist, a mechanical necromancer, a Frankenstein*

Cold War resonances are particularly evident in the attitude of Wolfe Mays (1912–2005), a philosopher at the University of Manchester at the time. Like Jefferson, Mays attended the 1949 Manchester seminars that preceded Turing's formulation of his test (cf. Chapter 4). In the summer of 1950, Mays reported in (2001), he was asked by Ryle to write a reply to Turing's paper to appear in the same October 1950 issue of *Mind*, but Ryle would have rejected his paper for being 'too polemical.' Later published (1952), Mays' paper offered a strong critique of Turing's paper. He rejected Turing's proposed association of the words 'machine' and 'thinking' and offered instead the words 'robot' and 'artifice:'

> [I]t may be necessary to introduce a new label to indicate a device which simulates overt human activities without at the same time duplicating our internal behaviour. The word is ready to hand and was coined by Karel Capek, we call them 'robots' … In this connection it might be a good thing to drop the word 'machine,' with its emotional overtones of clanging metal, and use some such neutral word as 'artifice.'[168]
>
> (Mays, 1952, p. 150)

Coined in the English translation of Capek's play *R.U.R. (Rossum's Universal Robots)* (1923), the word 'robot' comes from the Czech *robotnik* (forced laborer) and was associated with determinism and national-state tyranny. Mays went on to complain about '[t]he paradoxical Frankenstein nature of the machine-mind' (1952, p. 150), implying to Turing the designation of a 'mechanical necromancer' (p. 153).

In line with Mays' connotations on Turing as Dr. Frankenstein—and it is worth noting at this point Hodges' simultaneous promotion of the three images under analysis (Sections 7.3.1–7.3.3)—, Hodges characterizes Turing's views:

> His ruthless, raw view of science was something that Lyn Newman again captured with an image of him as 'the Alchemist' of the seventeenth century or before – recalling a time when science was not shrouded in titles and patronage and respectability, but was nakedly dangerous. There was a Shelley in him, but there was also a Frankenstein — the proud irresponsibility of pure science, concentrated in a single person.
>
> (Hodges, 1983, p. 521)

Hodges evokes Lyn Irvine's portrayal (1959) of Turing.[169] However, according to her, Turing 'certainly had less of the eighteenth and nineteenth centuries in him than most of his contemporaries.' And yet Hodges links Turing to Frankenstein, whose figure is surely more representative of scientists since the eighteenth-century Age of Enlightenment. Further, Hodges skipped over Irvine's clear note: 'His mother and his housemaster, of one mind about him throughout, saved Alan from what threatened to be a career of scientific pranks.' Nor does Hodges seem to have fully appreciated Irvine's fabulous account of the tenderness in Turing's eyes:

> ... his eyes ..., blue to the brightness and richness of stained glass. They sometimes passed unnoticed at first; he had a way of keeping them to himself, and there was also so much that was curious and interesting about his appearance to distract the attention. But once he had looked directly and earnestly at his companion, in the confidence of friendly talk, his eyes could never again be missed. Such candour and comprehension looked from them, something so civilized that one hardly dared to breathe. Being so far beyond words and acts, that glance seemed also beyond humanity.
>
> (Irvine, 1959, p. xxi)

Whether by what he left out or by what he emphasized, Hodges may have further stimulated the image of Turing as a Frankenstein that Mays had pushed through. Although Hodges supported Jefferson's image of Turing as a scientific Shelley, observing Turing's irreverence for patronage and institutional power, he later departed to some extent from Jefferson, who had used the word 'non-conformist' (see note 162) rather than 'irresponsible.' Hodges has created a multifaceted profile of Turing, 'the enigma,' based on his extensive research and the accounts of many of Turing's contemporaries. One of them, Turing's contemporary and close colleague Donald Michie (1984), acknowledges this in his review of Hodges: 'A single scene may yield many views—as many as there are observers.' Michie also noted that, although Hodges is a mathematician, his skill in writing his Turing biography 'is that of the novelist and the dramatist.'

Following along with Hodges' work, the image of Turing as Dr. Frankenstein has reappeared in non-expert readings of Turing, both in science and fiction. For example, Patrick Hayes and Kenneth Ford (1995) presented an influential critique of Turing's imitation test. Alluding to Frankenstein, they urged their colleagues to abandon the goal of creating an 'artificial human' and thereby free themselves from 'Turing's ghost.' Their commentary makes it sound as if Turing, unlike subsequent generations of AI scientists, was irresponsible.

Meanwhile, in fiction, Ian McEwan's recent novel *Machines Like Me* (2019) portrays Turing living beyond his 42nd year to lead the development of embodied artificial humans with artificial skin and a host of other features that make them resemble real humans. It makes it seem as if Turing's ambition was the synthesis of a human-like artificial being. Again, this is reminiscent of Mary Shelley's novel, which explores Frankenstein's drive to control nature and circumvent its processes in the laboratory.

Overall, descriptions of Turing as a dystopian actor, a mechanical necromancer like Frankenstein, seem intriguing against the backdrop of Turing's 1951 statements. I will address them in light of Turing's conception of an intelligent machine (Section 7.7).

7.4   The ironic Turing

Turing's sense of humor is highlighted in the accounts of his friends (Turing, 1959).[170] But what kind of humor did Turing exhibit and what can be drawn from it?

7.4.1   *Probably he did not mean this to be taken too seriously...*

Attention to Turing's use of humor, almost as a means of distracting from the seriousness of his views, goes back to his own time. A paradigmatic example can be found in the wake of the June 1949 polemic between Turing and Jefferson over the possibility of intelligent machines. As Chapter 4 shows (Section 4.6), *The Times* highlighted Jefferson's strong arguments, claiming a myriad of things that machines would never be able to do, most notably 'write a sonnet or compose a concerto because of thoughts and emotions felt.' And the next day, the reporter managed to get a reply from Turing, speaking for the Computing Machine Laboratory at Manchester University. Turing's ironic reply about sonnet-writing machines produced many reactions sent to the newspaper's correspondence.

For example, Hodges found that Dom Illtyd Trethowan, a Benedictine intellectual from Downside Abbey, was eager to respond to Turing's ironic words about sonnet-writing machines that appeared in *The Times*. In a letter to the newspaper, the monk addressed the 'responsible scientists' whom he urged 'to be quick to dissociate themselves' from Turing's research program.[171] (Again, it is worth noting that Hodges' reference to Turing's 'irresponsibility' echoed a committed historical actor.) Trethowan went on to warn that '[e]ven our dialectical materialists would feel necessitated to guard themselves, like Butler's Erewhonians, against the possible hostility of the machines.' He noted the institutional context of the computing projects in the UK, and encouraged those who saw human beings as 'free' persons to ask themselves 'how far Mr. Turing's opinions are shared, or may come to be shared, by the rulers of our country.'

Two weeks after Turing's words appeared on the media, the *British Medical Journal* published Jefferson's Lister Oration and an accompanying editorial with this rejoinder:

> Mr. A. W. [*sic*] Turing, who is one of the mathematicians in charge of the Manchester 'mechanical brain,' said in an interview with *The Times* (June 11) that he did not exclude the possibility that a machine might produce a sonnet, though it might require another machine to appreciate it. Probably he did not mean this to be taken too seriously ...
>
> (BMJ, 1949, p. 1129)

The same question arises again: how serious was Turing?

Max Newman, the sponsor of Turing's academic position as a Reader in the Mathematics Department at the University of Manchester, regretted the June 1949 polemic and tried to alleviate the social pressure with a clarifying note that appeared in the same issue of the BMJ (Newman, 1949). He emphasized that digital computing machines could handle a variety of computing problems and observed:

> The first question that will have to be asked is not 'Can all *kinds* of thought, logical, poetical, reflective, be imitated by machines?' but 'Can *anything* that can be called "thought" be so imitated and, if so, how much?'
>
> (Newman, 1949, p. 1133, his emphasis)

Following this moderate note, Newman remarked: 'The most promising line here will be to work within mathematics itself.' According to Newman's wife Lyn Irvine (1949), his letter was an attempt to 'clear things up.' But Turing's sonnet-writing machine remark suggests that he drew no line between Newman's two questions, universal and existential. There must have been something of value to Turing that he felt was at stake. He was undeterred by the public reaction to his views.

7.4.2 *Humor with intellectual integrity*

An example of Turing's willingness to take his progressivism a long way is his attitude toward the police when he was charged with sexual offenses. In January 1952, a case was formed against him under 'Gross Indecency contrary to Section 11 of the Criminal Law Amendment Act 1885.' According to Hodges (1983), the detectives reported: 'He was a real convert ... he really believed he was doing the right thing' (p. 457).

Overall, an informative note on the nature of Turing's humor was given by Turing's fellow mathematician and contemporary at King's College, Denis Williams:

> In intellectual, as in other matters, it was essential to him that everything should ring true. [...It] seems to me precisely this complete intellectual integrity, which, combined with his other gifts, made it reasonable to expect that he would produce results of fundamental importance in his own field. Alan had a delightful sense of humour. He enjoyed elaborating fantastic projects, such as a scheme for faking prehistoric cave paintings, in mock-serious detail, or bringing an over-serious discussion down to earth with a quick colloquial turn of phrase. With him jest and earnestness were often closely intermingled.
>
> (Turing, 1959, p. 91)

Now, if humor and intellectual integrity went hand in hand for Turing, how can we make sense of his ironic statements?

7.4.3　Irony as a method

Against great, as Turing put it, 'intellectual' and 'emotional' opposition, he made quite extensive use of irony. Sensing that he was not being properly heard, Turing found his way to respond. John V. Price's interpretation of David Hume's irony can be helpful to our understanding of Turing's:

> When a man was under the intellectual and cultural pressure which Hume experienced he could not respond easily by denunciations, by shouting, or by threats. As a civilized man, Hume would not have responded that way under any circumstance. His method of dealing with those who would persecute him or ostracize him simply because of his religious or philosophical or moral opinions was subtle and effective. Irony gave him a method of operating in a world that found his ideas both strange and shocking: strange because most people were simply unable to handle them, shocking because his scepticism dared to attack the citadel of religion. New ways of thinking about man's place in nature, especially if they do not reassure one's blind faith, are often difficult ... to tolerate. Irony could at least create artificial tolerance.
>
> (Price, 1965, pp. 4–5)

This proves to be very enlightening when applied to Turing. His 'new ways of thinking about man's place in nature' were 'often difficult to tolerate' indeed. This parallel suggests that Turing's words should neither be understood literally nor dismissed as plain mockery. Rather, his irony can be understood as a clever form of communication in an environment that was not receptive to him. It can be noted that Turing subtly used irony to make a point—e.g. in his comment to *The Times* about sonnet-writing machines— by imitating people's language in ways that were aimed at exposing their vices or stupidity. This is different from mere parody, gentle or otherwise, or imitation for (philosophically empty) comic effect. I suggest that Turing's irony can be best understood as satire, i.e. irony with a point. Under this interpretation, we can now examine the point that Turing was trying to make.

7.5   Turing's Promethean irreverence

Turing's move in 1951 was not a thoughtless, isolated step. Rather, the same argument is consistently present in Turing's communications and writings every year from 1947 to 1951 in connection with his conception of an intelligent machine. (We have seen his reply to *The Times* in June 1949, and will soon see his lecture to the London Mathematical Society in February 1947.) I will now show that in 1948 and 1950, through his systematic use of irony, he consistently articulated what he called—but denied being guilty of—his 'Promethean irreverence,' consisting of what can be understood as an ontological and ethical argument against two objections to the possibility of intelligent machines.

7.5.1   *The 1948 objections (a) and (b) to intelligent machines*

Turing wrote his last NPL report on 'Intelligent Machinery' in 1948 while on leave from the NPL to which he would never return. I want to emphasize that the first two of the five objections articulated there have the same structure as his 1951 BBC radio broadcasts — namely, they address what can be seen as an ontological and an ethical question, respectively:

> (a) An unwillingness to admit the possibility that mankind can have any rivals in intellectual power. This occurs as much amongst intellectual people as amongst others: they have more to lose. Those who admit the possibility all agree that its realization would be very disagreeable. The same situation arises in connection with the possibility of our being superseded by some other animal species. This is almost as disagreeable and its theoretical possibility is indisputable.
> (b) A religious belief that any attempt to construct such [intelligent] machines is a sort of Promethean irreverence.
>
> (Turing, 1948, p. 410)

While objection (a) deals with the ontological question of whether intelligent machines could ever exist, objection (b) deals with the ethical question of whether one should ever build such machines. In objection (a), Turing refers to a kind of conflict of interest that he saw in the position of 'intellectual people.' In objection (b), he responds to the charge that any attempt to build an electronic brain would be 'a sort of Promethean irreverence.' Turing answered both objections:

> The objections (a) and (b), being purely emotional, do not really need to be refuted. If one feels it necessary to refute them there is little to be said that could hope to prevail, though the actual production of the machines would probably have some effect.
>
> (Turing, 1948, p. 410)

This suggests that Turing relied on the actual existence of intelligent machines as the best approach to addressing objections (a) and (b). Note also that objection (a), Turing's first formulation of an objection to intelligent machines, marks the moment when he first expressed the core idea of 'The Book of Machines' in Butler's *Erewhon*, though without citing it on that occasion. The Butlerian idea in objection (a) is, essentially, the view of machines as a species (in Butler's terms, 'the mechanical kingdom'). Through evolution, they could in principle rival and surpass humans in intellectual power. This was a corollary of Butler's critique of Charles Darwin, as we will soon see (§7.6). This idea would recur in Turing's work until 1951, as we have seen from his statements in the two BBC radio broadcasts (§7.1).

7.5.2 *The 1950 reformulation as objections (1) and (2)*

In (1950), Turing changed the order of the two 'purely emotional' objections, (a) and (b), and named them 'the theological objection' and 'the "heads in the sand" objection:'[172]

> (1) *The Theological Objection.* Thinking is a function of man's immortal soul. God has given an immortal soul to every man and woman, but not to any other animal or to machines. Hence no animal or machine can think.
> … In attempting to construct such [intelligent] machines we should not be irreverently usurping His [God's] power of creating souls, any more than we are in the procreation of children: rather we are, in either case, instruments of His will providing mansions for the souls that He creates.

> (2) *The 'Heads in the Sand' Objection.* "The consequences of machines thinking would be too dreadful. Let us hope and believe that they cannot do so."

... We like to believe that Man is in some subtle way superior to the rest of creation. It is best if he can be shown to be *necessarily* superior, for then there is no danger of him losing his commanding position. The popularity of the theological argument is clearly connected with this feeling. It is likely to be quite strong in intellectual people, since they value the power of thinking more highly than others, and are more inclined to base their belief in the superiority of Man on this power.

(Turing, 1950, pp. 443–444, his emphasis)

Once again, Turing linked the two objections by referring to the first in his discussion of the second. In objection (1), Turing referred to the image of usurping God's power to create souls, and thus indirectly to the myth of Prometheus, and again in connection with religion, now satirized. He suggests that the ethical claim against the project is based on unfounded ontological assumptions about distinctions between humans and other intelligent animals or machines. In objection (2), Turing again focuses on 'intellectual people' and refers to 'Man' and his unwillingness to lose 'his commanding position.'

Turing finished his 1950 paper hoping that machines would be intellectually competitive 'with men in all purely intellectual fields' and called for action to make it happen, addressing both questions of ontology and ethics at once. Turing's argument in response to these questions will become clearer through his conception of an intelligent machine (Section 7.6) and its related social implications (Section 7.8).

7.6   Satiric novels and Turing's conception of an intelligent machine

Turing had seen engaging models of satire since his youth, as his mother Mrs. Sara Turing reported (1959, p. 108): 'In his late teens he read a certain amount of fiction.' Specifically, she noted: 'He had a particular fondness for The Pickwick Papers ... and Samuel Butler's Erewhon. This last possibly set him to think about the construction of an actual intelligent machine.' Charles Dickens' *The Pickwick Papers* (1836) and Samuel Butler's *Erewhon* (1872) can be quite informative not only about Turing's satirical style but also about his conception of an intelligent machine. This will further support the claim that Turing used irony with a point, that satire is deeply integrated into his science. 'With him,' as Denis Williams observed, 'jest and earnestness were often closely intermingled.'

7.6.1   *Turing's Dickensian satire of the universal machine as an image of the mind*

Charles Dickens (1812–1870) was an English writer and critic of Victorian society. Dickens made extensive use of the word 'mechanical,' satirizing unreflective behavior (e.g.): 'Mr. Winkle, being half asleep, obeyed the command mechanically, opened the door a little, and peeped out;' or "'I'm a-comin', sir,'" replied Mr. Weller, mechanically following his master' (Dickens, 1836). Jon Agar (2003) notes Dickens' specific innovation relative to the mechanical metaphor (p. 64). He refers, for example, to Dickens' *Little Dorrit* (1857) as a satire of the increasingly rule-bound British Civil Service: 'Because the Circumlocution Office went on mechanically, every day, keeping this wonderful, all-sufficient wheel of statesmanship, How not to do it, in motion' (p. 76).[173]

In his seminal 'On Computable Numbers' (1936), Turing only used the word 'mechanical' in his final move to refute Hilbert's program for the complete mechanization of mathematics:

> We are now in a position to show that [Hilbert's] Entscheidungsproblem cannot be solved. Let us suppose the contrary. Then there is a general (mechanical) process for determining ...
>
> (Turing, 1936, p. 262)

Following the war, the universal (Turing) machine would be detached from its original 1936 context and recast in the UK, with Turing's approval, as a positive concept in connection with computer technology. This recasting appears notably in the documentation of the UK's national computing project (NPL, 1946). In Turing's obituary, Max Newman (1955) did consider that the emerging technology of 'automatic' computing machines 'were in principle realizations of the "universal machine."' However, he did note that Turing initially conceived the concept 'for the purpose of a logical argument' (p. 254). Turing's 1936 concept of the universal machine served an *instrumental* use as part of a negative proof: it was conceived to refute Hilbert's program for the complete mechanization of mathematics, and certainly not to serve as a general model for the human mind.

Turing (1950) Turing did consider the possibility that the whole mind might be 'mechanical' (p. 455), but this use of the word in the context of his discussion of human and machine intelligence was different from that of (1936), which focused on the activity of human clerks. Since Turing's machines, the discussion of mind and mechanism has largely centered on arguments for and against seeing the human mind as a machine (Putnam, 1960; Lucas, 1961).[174] However, should an adult human mind be understood in this way, Turing satirized in 1948 in Dickensian style, emphasizing the unthinking following of orders:

> This would mean that the adult will obey orders given in appropriate language, even if they were very complicated; he would have no common sense, and would obey the most ridiculous orders unflinchingly. When

all his orders had been fulfilled he would sink into a comatose state or perhaps obey some standing order, such as eating. Creatures not unlike this can really be found, but most people behave quite differently under many circumstances.

(Turing, 1948, p. 424)

Turing's intelligent machines, rather than mechanical devices in the Dickensian sense of unthinkingly following orders and rules, were imagined to be capable of discussing a literary interpretation of Mr. Pickwick (Turing, 1950, pp. 446–447). They were based on his concept of a 'learning' machine (1950), developed from his earlier concept of the 'unorganized' machine (1948), as opposed to the 'universal' machine, which strictly follows instructions.[175]

7.6.2   *Turing's evolutionary machines and Samuel Butler*

Turing's learning machines would acquire intelligence in childhood through environmental interference in the form of an external educator:

There is an obvious connection between this process and evolution, by the identifications

Structure of the child machine = Hereditary material
Changes of the child machine = Mutations
Natural selection = Judgment of the experimenter

One may hope, however, that this process *will be more expeditious than evolution*. The survival of the fittest is a slow method for measuring advantages. *The experimenter*, by the exercise of intelligence, should be able to speed it up.

(Turing, 1950, p. 456, my emphasis).

Here is Turing's defense of the possibility of rapid cultural evolution of intelligent machines on top of the universal machine (fixed hardware). In Turing's bibliography (1950) appears 'The Book of the Machines,' which is a major part of *Erewhon* (Butler, 1872). The latter is the most famous of Butler's novels and is widely regarded as a satire on the values and manners of Victorian England.[176] Like Dickens, Samuel Butler (1835–1902) was a Victorian novelist and social critic. He respected no boundaries in his attacks on the social, religious, and scientific establishments, and he used irony to express his deeply held views. In his literary career, Butler presented a fierce critique of Charles Darwin's theory of evolution by natural selection.[177] 'The Book of the Machines' had appeared in an earlier version in 1863 as 'Darwin among the Machines.' As a satire of Darwin's theory, it focuses on the systemic human displacement that is caused by the evolution of machines, which emerges as the central theme of Turing's heretical theory of intelligent machines. The central question Butler poses to Darwin is that either Darwin had deliberately ignored the problem of the creation of the original germ of life from which all other living things evolved, or he intended it to be assumed that life was in fact indistinguishable from matter and had somehow evolved from matter. Butler explores the second alternative, and Turing follows him in this. Here is Butler in *Unconscious Memory*:[178]

> I first asked myself whether life might not, after all, resolve itself into the complexity of arrangement of an inconceivably intricate mechanism. Kittens think our shoe-strings are alive when they see us lacing them, because they see the tag at the end jump about without understanding all the ins and outs of how it comes to do so ... Suppose the toy more complex still, so that it might run a few yards, stop, and run on again without an additional winding up; and suppose it so constructed that it could *imitate* eating and drinking, and could make as though the mouse were cleaning its face with its paws. Should we not at first be taken in ourselves, and assume the presence of the remaining facts of life, though in reality they were not there?
>
> (Butler, 1880, ch. 2, my emphasis)

Compare Turing's 1950 imitation game, proposed as a criterion for the presence of an intelligent machine. Butler continued:

> If, then, men were not really alive after all, but were only machines of so complicated a make that it was less trouble to us to cut the difficulty and say that that kind of mechanism was 'being alive' ...?
>
> (Butler, 1880, ch. 2)

Compare Turing (1948), referring to chess-playing machines (p. 412): 'Playing against such a machine gives a definite feeling that one is pitting one's wits against something alive.' Along these lines, Butler asked:

> ... why should not machines ultimately become as complicated as we are, or at any rate complicated enough to be called living, and to be indeed as living as it was in the nature of anything at all to be? If it was only a case of their becoming more complicated, we were certainly doing our best to make them so.
>
> (Butler, 1880, ch. 2)

This is the essence of Butler's view of the evolution of machines seen as a species. Butler added that he eventually realized that:

> ... this view comes to much the same as denying that there are such qualities as life and consciousness at all, and that this, again, works round to the assertion of their omnipresence in every molecule of matter, inasmuch as it destroys the separation between the organic and inorganic, and maintains that whatever the organic is the inorganic is also.
>
> (Butler, 1880, ch. 2)

'And it is to this second conclusion that Butler does indeed come,' explained the literary critic and Butler's biographer Nicholas Furbank (1946, p. 59). 'If it is to be either all mind or no mind,' Furbank added, 'we might as well plump for all mind.' Furbank was nothing less than one of Turing's two closest friends, whom Turing chose as his will executor. Furbank's biography of Butler, which develops the above quotes, was published by the Cambridge University Press in 1948, the same year that Turing and Furbank met for the first time at the end of the summer. Butler, as we have seen, is an author who, according to Turing's mother, had captured the imagination of the young Turing. Now, compare Butler's view just quoted with Turing (1950): 'I do not wish to give the impression that I think there is no mystery about consciousness' (p. 447). He added: 'But I do not think these mysteries necessarily need to be solved before we can answer the question with which we are concerned in this paper' (can machines think?).

Given Butler's influence on Turing, it should not be surprising that the intelligent machines that Turing envisioned could possibly acquire the capabilities of: 'be kind, resourceful, beautiful, friendly, have initiative, have a sense of humour, tell right from wrong, make mistakes, fall in love, enjoy strawberries and cream, make some one fall in love with it, learn from experience, use words properly, be the subject of its own thought, have as much diversity of behaviour as a man, do something really new' (p. 447). After arguing this, Turing wrote: 'These are possibilities of the near future, rather than Utopian dreams' (p. 449).

This indicates that Turing thought his project was more realistic than utopian, but more generally it indicates that he presented the project in the optimistic frame of a utopia rather than in the pessimistic frame of a dystopia.

7.7 Turing as Dr. Frankenstein?

In her famous *Frankenstein, or the Modern Prometheus* (1818), Mary Shelley (1797-1851) presented the character of Dr. Victor Frankenstein, a scientist who injects the spark of life into an otherwise inert body, only later to be horrified by his creature. There is a sense of despair in Mary Shelley's narrative as we see Frankenstein pushing his project forward unreflectively. Mary Shelley's concern appears, for instance, in the creature murmuring:

> Like Adam, I was created apparently united by no link to any other being in existence; but his state was far different from mine in every other respect. He had come forth from the hands of God a perfect creature, happy and prosperous, guarded by the especial care of his Creator; he was allowed to converse with, and acquire knowledge from beings of a superior nature: but I was wretched, helpless, and alone.
>
> (Shelley, 1818, ch. 7)

Abandoned by Frankenstein, the creature lacked protection, assistance, and education in his (cultural) infancy. Later in her literary career, Shelley would refer to Evangelical discourses elevating not God but the mother as the theological and moral center of the household (Airey, 2019).

It is hard to blame Turing for being unreflective about the future of machines in society. For example, he wrote (1950): 'I believe further that no useful purpose is served by concealing these beliefs [about the feasibility of intelligent machinery]' (p. 442). Drew McDermott (2007) acknowledged Turing's contributions to this discussion, writing that 'Turing was his own Mary Shelley.' McDermott thus saw Turing as a reflective, dystopian thinker, and this is somewhat at odds with Hayes and Ford's view of him as a dystopian actor. And yet there seems to be less of the sense of despair typical of dystopian narratives in Turing's remarks, and more of a sense of a utopian satire. Turing seems to have been far ahead of his contemporaries in his foresight of the unimagined possibilities of digital computers. Further, Turing did not believe that humans were superior beings—according to Descartes (1637, Part VI), 'the lords and possessors of nature' —, quite the opposite. Turing's refusal to see humans as masters of nature is at odds with Hodges's suggestion that Turing embraced arrogant scientism along the lines of Mary Shelley's Frankenstein.

Besides, in Mary Shelley's novel, Frankenstein's drive to control nature takes a specific form in his collapse of the natural and the artificial (synthetic). As noted above, Hayes and Ford pleaded with their colleagues to abandon the goal of creating an 'artificial human' (1995), and in McEwan's novel (2019), Turing's work leads to the synthesis of embodied artificial humans with artificial skin and a host of other features that make them look like real humans. However, after one of Turing's outrageous remarks in 1951, he said:

> It is customary, in a talk or article on this subject, to offer a grain of comfort, in the form of a statement that some particularly human characteristic could never be imitated by a machine. It might for instance be

said that no machine could write good English, or that it could not be influenced by sex-appeal or smoke a pipe. I cannot offer any such comfort, for I believe that no such bounds can be set. But I certainly hope and believe that no great efforts will be put into making machines with the most distinctively human, but non-intellectual characteristics such as the shape of the human body; it appears to me to be quite futile to make such attempts and their results would have something like the unpleasant quality of artificial flowers. Attempts to produce a thinking machine seem to me to be in a different category.

(Turing, 1951a, p. 486)

This suggests that Turing did not appreciate the prospect of making an artificial creature that resembled the human body and even thought this to be futile and bound to give unpleasant results. Turing's reference to the 'unpleasant quality of artificial flowers' underlines a distinction between the natural and the artificial, where 'artificial' means synthesized as opposed to evolved, raised or grown like in the processes of nature.[179]

Turing envisioned raising his child machines as analogous to raising a human child. In his report 'Intelligent Machinery' (1948), Turing observed that the education of a human being may take 'twenty years or more,' emphasizing the importance of social contact at least until 'university graduation' (p. 421). He pointed out that 'the isolated man does not develop any intellectual power' (p. 431). In (1950), Turing satirized that a machine could be tutored and could even go to school if it was not for the fact that 'other children' would make 'excessive fun of it' (p. 456). Susan Sterrett points out (2012, p. 709) that Turing's insight that social and cultural interference in child rearing is crucial to the development of intelligence is not outdated in the light of research findings in cultural anthropology. Moreover, Turing did not consider the possibility of embodied, cyborg-like machines. He even focused his analogy between humans and machines on the case of 'Miss Helen Keller,' who was able to educate herself despite her physical disabilities. Turing envisioned intelligent machines with limited agency in the world, functioning more as intellectual and educational tools. This is different from today's AI, which can largely be regarded as agency without true intelligence (Floridi, 2023).

In this original sense, Turing imagined a society permeated by intelligent machines that would develop a subjectivity based on their own individual experiences, as opposed to a gigantic database of collective human experiences. Their behavior would go beyond a mechanical reproduction of what they were taught, including reflective responses similar to those of human children. Turing was remembered as saying: 'proud owners [would] say "My machine" (instead of "My little boy") "said such a funny thing this morning"' (Newman, 1955, p. 255). Although the latter may already be a reality with present-day AI, note that the owner and educator of an intelligent machine in Turing's imagined future is a person, not a nation-state or a large corporation.

Overall, given Turing's focus on the education of his 'child' machines (Sterrett, 2012, 2017), Mary Shelley's concern about scientists like Dr. Frankenstein abandoning their creature without any cultural upbringing could hardly apply to Turing.

7.8   Turing's intelligent machine utopia

The word 'utopia' comes from the Greek 'topos,' meaning 'place' or 'where,' and 'u' from the prefix 'ou,' meaning 'no' or 'not,' and has come to refer to an ultimately good but non-existent place. For this reason, 'to call something "utopian" has, from very early on, been a way of dismissing it as unrealistic' (Sargent, 2010, pp. 14–15). A utopian frame of mind arises as a result of bad times. The experience of bad times produces visions of a utopian future in which the evils of society have been eliminated, replaced, or transcended, which is usually for the benefit of all humanity.

Frank and Fritzie Manuel's large study of utopian thought in the West connects another important element, which is technology:

> Every utopia, rooted as it is in time and place, is bound to reproduce the stage scenery of its particular world as well as its preoccupations with contemporary social problems. Utopias … avail themselves of the existing equipment of a society, perhaps its most advanced models, prettified and rearranged. Often a utopian foresees the later evolution and consequences of technological development already present in an embryonic state.
>
> (Manuel and Manuel, 1979, p. 23)

The connection with Turing's perspective on the future of early digital computers in society, as opposed to the most obvious scientific and military applications of computing in the early 1950s, seems striking.

What specific evils would have led Turing into a utopian frame of mind? We have seen that Turing avoided addressing issues explicitly, preferring to do so in passing, satirically (Section 7.4). However, through his remarkable use of irony, he provided several clues to his concerns about the time and place in which he lived. Once collected and connected, they may be revealing.

7.8.1   *Turing's critique of habits and customs*

Hodges seems to side with Jefferson in interpreting Turing's philosophy of mind as reductionist, writing, for example (1983): 'he introduced the idea of an operational definition of 'thinking' or 'intelligence' or 'consciousness' by means of a sexual guessing game' (p. 415). However, this view is at odds with the fact that Turing explicitly rejected the need to provide such a definition.

In fact, by introducing the imitation game, which requires a machine to imitate a woman in one version and a man in another, Turing moved the discussion of thinking and intelligence away from a standard analytical approach to philosophy. Hodges seems not to have appreciated Turing's imitation game as irony with a point. He interpreted Turing's man-imitates-woman variant of the game as 'a red herring, and one of the few passages of the paper that was not expressed with perfect lucidity' (1983, p. 415).

It is worth pausing for a moment to appreciate the point of Turing's imitation game. Chapter 5 reconstructs how Turing methodically varied the settings of the game, using case and control variants to experiment with the question: can player A successfully imitate stereotypes associated with player B's type, despite their physical differences? Rather than promoting a universal concept of 'human intelligence,' Turing's 1950 argument can be understood as Sterrett (2000) has suggested, as critically addressing stereotypes of intelligent *v.* mechanical, female *v.* male, human *v.* machine, natural *v.* artificial. Juliet Floyd (2017) adds strength to this interpretation, noting that Turing 'saw the difference in levels and types as a complex series of systematizations sensitive to everyday "phraseology" and common sense, not a divide of principle' (p. 142). This, Floyd adds, 'was because he always saw "types" or "levels" as lying on an evolving continuum, shaped by practical aspects, the user end, and mathematics.'[180] On gender specifically, Turing understood as early as 1950 that gendered behavior is taught and learned in the child's upbringing (see Sterrett, 2000, 2012, 2017). Turing was probably also responding to a thought experiment of Jefferson's which suggested that gendered behavior is causally determined by the physiology of male and female sex hormones (see Chapter 5).

Throughout his ironic exposition of the imitation game, Turing implicitly argued that intelligent machines could cause intellectuals to confront their own prejudices. He satirized the 'theological' objection and the 'arguments from various disabilities:'

Thinking is a function of man's immortal soul. God has given an immortal soul to every man and woman, but not to any other animal or to machines ... I am unable to accept any part of this ... The arbitrary character of the orthodox view becomes clearer if we consider how it might appear to a member of some other religious community.

(Turing, 1950, p. 443)

The works and customs of mankind do not seem to be very suitable material to which to apply scientific induction. A very large part of space-time must be investigated, if reliable results are to be obtained. Otherwise we may (as most English children do) decide that everybody speaks English, and that it is silly to learn French.

(Turing, 1950, p. 448)

What is important about this disability [being unable to enjoy strawberries and cream] is that it contributes to some of the other disabilities, *e.g.* to the difficulty of the same kind of friendliness occurring between man and machine as between white man and white man, or between black man and black man.

(Turing, 1950, p. 448)

With his peculiar touch of irony, Turing addressed chauvinisms of religion and ethnicity, nationality, and race. These added to his strong critique of species chauvinism, which shifted the burden of proof and invited humanity to look at itself.

Beyond habits and customs, Turing's irony addressed social and institutional structures, especially the division of labor and the role of intellectuals. By doing so, he may be seen as echoing the unheard voices from the relatively progressive environment of Bletchley Park, as Hodges (1983) nicely suggests, where 'those excluded from participation in peace— ordinary men, the young, and even women,' all had played crucial parts (p. 311).

7.8.2   *Turing's critique of social and institutional structures*

References to Frankenstein appeared in the English press in November 1946,[181] in connection with the advent of modern computing machines and the growing buzz about scientists building an 'electronic brain.' This term, Hodges found through his extensive research (1983, p. 347), appeared in a speech by Louis Mountbatten (1946) to the British Institution of Radio Engineers on October 31, 1946, and the next day in an article about it in *The Times*. The British statesman spoke of a 'revolution of the mind' and of chess-playing machines. He relied on information obtained from the National Physical Laboratory (NPL),[182] in particular Turing's earliest postwar projections about the future of computing and the possibility of teaching machines to play chess, which Turing had presented in his report (1945, p. 389) to the Executive Committee.

It is known from Donald Bayley, who worked with Turing from 1943 to 1946, that Turing had spoken at the end of the war of his intention 'to build a brain' (Hodges, 1983; Sykes, 1992, p. 290; 25–27'). He joined the NPL in October 1945 to pursue this project, although he knew that the digital computing projects in the US and the UK were an outgrowth of World War II. Turing sought to shift their use from mechanizing computation, primarily for warfare,[183] to supporting fundamental studies of nature and mathematics. Just after the media frenzy in early November, 1946, Turing wrote to Ross Ashby (1946): 'I am most interested in the possibility of producing models of the action of the brain than in practical applications to computing.'

But Turing's interests were in sharp contrast to the NPL leadership's narrative, as the letter from Darwin, the NPL director, to *The Times*, shows. Darwin intended to clarify the NPL's official position on the 'electronic brain' polemic:

> In popular language the word 'brain' is associated with the higher realms of the intellect, but in fact a very great part of the brain is an unconscious automatic machine producing precise and sometimes very complicated reactions to stimuli. This is the only part of the brain we may aspire to imitate. The new machines will in no way replace thought, but rather they will increase the need for it ...[184]

From a social perspective, it is worth noting one consequence of NPL's statement: it distinguishes tasks and jobs that should be automated from those that should not. Turing did not accept this division. Intellectual jobs should experience the same displacement as labor-intensive jobs.

Before Darwin, as Chapter 4 shows (Section 4.4), Douglas Hartree was the first intellectual to respond publicly to Mountbatten's speech, and thus indirectly to Turing. In fact, Darwin's letter to *The Times* was an official endorsement of a letter Hartree had written to *The Times* a few days earlier, attacking the use of the term 'electronic brain' and claiming that computing machines 'can only do precisely what they are instructed to do.'[185] Hartree made the same point in his inaugural lecture at Cambridge University (1947), where he described the ENIAC core unit—the so-called 'master programmer'— as a device that allowed the 'automatic control' of computation and 'endows the machine with judgement in a restricted sense,' since the machine 'can only do strictly and precisely what it is told to do' (pp. 20–21).[186]

A few months after Hartree and Darwin's public letters in November 1946, on February 20, 1947, Turing gave a talk to the London Mathematical Society about the machine building project at the NPL. At the end of his talk, he made the master-slave dichotomy a prevailing subject:

> It has been said that computing machines can only carry out the processes that they are instructed to do. This is certainly true in the sense that ... the intention in constructing these machines in the first instance is to treat them as slaves, giving them only jobs which have been thought out in detail, jobs such that the user of the machine fully understands what in principle is going on all the time.
>
> (Turing, 1947, pp. 392–393)

Turing resumed observing that '[u]p till the present machines have only been used in this way,' and asked: 'But is it necessary that they should always be used in such a manner?' (p. 393).[187] Turing's plea to liberate 'machines' from slavery followed his questioning of the ethics of master programmers:

> Roughly speaking those who work in connection with the [National Physical Laboratory Automatic Computing Engine] will be divided into its masters and its servants. Its masters will plan out instruction tables for it, thinking up deeper and deeper ways of using it. Its servants will feed it with cards as it calls for them ... As time goes on the calculator itself will take over the functions both of masters and of servants.
>
> (Turing, 1947, p. 392)

The displacement of both masters and servants by the machine was thus implied early on in Turing's postwar communications. He resumed this and casually revealed that the masters-servants division, in this context, also corresponded to a gender division: 'One might for instance provide curve followers to enable data to be taken direct from curves instead of having girls read off values and punch them on cards.' Here, Turing's earlier references to 'Man' as a masculine generic were in fact materially marked by the male gender,[188] although his focus in this passage is on the division of labor and the power imbalance between 'intellectual' and non-intellectual, 'mechanical' work that was evident at Bletchley Park. Turing thus emphasized his concern that the 'masters' would perceive intelligent machines as a threat to their dominant position:

> The masters are liable to get replaced because as soon as any technique becomes at all stereotyped it becomes possible to devise a system of instruction tables which will enable the electronic computer to do it for itself. It may happen however that the masters will refuse to do this. They may be unwilling to let their jobs be stolen from them in this way. In that case they would surround the whole of their work with mystery and make excuses, couched in well chosen gibberish, whenever any dangerous suggestions were made. I think that a reaction of this kind is

a very real danger. This topic naturally leads to the question as to how far it is possible in principle for a computing machine to simulate human activities.

(Turing, 1947, p. 392)

If machines displaced the lower classes of workers, but not the higher ones, that was the real danger, according to Turing.[189] Some of the 'masters' he was addressing, intellectuals in positions of power, must have been in the audience of this lecture when Turing said to their faces that they would try to repress intelligent machines. Feeling threatened by 'any dangerous suggestions' the machines might make, the intellectuals would 'surround the whole of their work with mystery and make excuses, couched in well chosen gibberish.'

Hodges (1983) linked Turing's remarks as a response to Darwin's letter to *The Times* quoted above, observing: 'To describe such careful and responsible statements [Darwin's] as 'gibberish' was not the most tactful policy' (p. 357). However, it can be noted that in 1947 Turing deliberately challenged Darwin, Hartree, and the NPL senior management, and possibly other mathematicians in the audience of his lecture. In 1948, Turing generalized his argument to 'intellectual people' and their 'unwillingness to admit the possibility that mankind can have any rivals in intellectual power.' In 1951 he addressed protectionist 'intellectuals' who would be 'afraid of being out of a job.'

It can be understood that it was less the 'human race' as a whole that Turing welcomed receiving a humiliating lesson from machines, but a particular class or group — chauvinists of all kinds, especially intellectuals in dominant positions. The intelligent machines Turing envisioned would be able, contrary to what Hartree expected, to do more than 'strictly and precisely' what they are told to do, and contrary to what Darwin expected, to 'imitate' not only the lower realms of the intellect but also the higher ones typically associated with 'thought.' Therefore, they would affect not only jobs that are considered lower, but also jobs that are considered higher, potentially challenging existing social and institutional structures and helping to democratize the distribution of power in society.

7.9 Conclusion

I suggest that Jefferson's image of Turing as a scientific Shelley is more profoundly correct than Hodges could present, given his concern to provide a multifaceted portrait of Turing.

Percy B. Shelley's masterpiece *Prometheus Unbound* (1820) is a four-act lyric drama based on the trilogy of *Prometheia* by the Greek playwright

Aeschylus. The classical trilogy concerns the torment of the Greek mytho-logical figure Prometheus, who defies the gods and gives fire to mankind, for which he is condemned to suffering and eternal punishment by the god Jupiter, a representation of order and power. The stolen fire here has the metaphorical meaning of intelligence, whose specific form varies with the versions of the myth (e.g., political intelligence in Plato's version of the story in *Protagoras*). In contrast to Aeschylus' version, Shelley presented a Prometheus who, with the help of his beloved Oceanid Asia, manages to free himself from Jupiter's oppression and start a bloodless revolution. In the words of Mary Shelley (1839), Percy 'followed certain classical author-ities in figuring … Prometheus as the regenerator, who, unable to bring mankind back to primitive innocence, used knowledge as a weapon to defeat evil, by leading mankind beyond the state wherein they are sinless through ignorance, to that in which they are virtuous through wisdom' (pp. 684–685). In Turing's utopia, child machines rely on learning and reason to bloodlessly confront the ruling elite intellectuals, just as Asia's supportive character mediates Prometheus' confrontation with the gods.

Exploring the first image of Turing, this chapter has examined Turing's irony in its social and historical context and followed the internal logic of his arguments to their logical conclusion. By doing so, it opens a new histo-riographical perspective on Turing's work. It is possible to see that his intel-ligent machine utopia is directed against social and institutional structures, chauvinistic views of society and nature, and intellectuals who might sacri-fice independent thought to maintain their power. Such intellectuals, Turing hoped, would eventually be surpassed by intelligent machines and trans-formed into ordinary people, as work once considered 'intellectual' would be transformed into non-intellectual, 'mechanical' work. Educated by indi-viduals, rather than nation-states or large corporations, his ever-learning child machines would outstrip powerful intellectuals and turn them into ordinary people, helping to distribute power more evenly in society. Tur-ing believed that the possibilities of the machines he envisioned were not utopian dreams.

Was Turing realistic? That may be seen as an open question. Recent sci-entific and technological advances, notably those demonstrated by a sys-tem called ChatGPT (Floridi, 2023), have made Turing's argument that machines may indeed outstrip humans in intelligence seem more realis-tic. At the same time, the power of machine intelligence seems increasingly concentrated in the hands of a few. The idea of ever-learning machines, whose intelligence would grow out of their individual experiences and help democratize power in society, may therefore still seem a distant reality.

8　Conclusion

This chapter concludes the book.

Concluding remarks

This book has presented a new interpretation and perspective on the Turing test and Turing's construction of machine intelligence, largely the result of using methods from the history of philosophy and the history and philosophy of science. With Turing's peculiar character, he was intensely witty and wholehearted in presenting his argument and at the same time intensely serious and profound in defending his controversial views.

More than seventy years later, as mentioned, the Turing test argument still seems to be received in extremes, either exaggerated by Whiggish excitement (the great foundation of artificial intelligence) or dismissed outright as philosophically naive. Meanwhile, machine intelligence continues to evolve and reach new milestones, albeit in some ways away from Turing's original views. As its impact on society grows, new opportunities arise to remember Turing and to reconsider what exactly he was trying to say.

I have examined Turing's test proposal in the light of its reception history, and in the context of conceptual chronologies of his communications and writings, and in its historical, social, and cultural context. On the one hand, Turing's original proposal of his test was found to be a thought experiment in the modern scientific tradition, proposed in the context of a controversy, and at odds with the long-held idea of a crucial experiment for artificial intelligence. On the other hand, it was found to be a sophisticated philosophical argument that the ideal of natural order by which we think about minds and machines would shift in the presence of more sophisticated machines — 'the time had come for philosophers and mathematicians and scientists to take seriously the fact that computers were not merely calculating engines but were capable of behaviour which must be accounted as intelligent,' or 'as we may say more briefly, if less accurately,' that machines can think.

This is a relatively modest first step in the context of a larger intellectual project about Turing's construction of machine intelligence and its relation to human intelligence, in its philosophical, historical, social, and cultural roots.

DOI: 10.4324/9781003300267-8

Notes

1 Comparably, the 12-old Turing once wrote: 'I always seem to want to make things from the thing that is commonest in nature and with the least waste in energy.' Turing to his parents from Hazelhurst, Sussex, on March 15, 1925. Archives Centre, King's College, Cambridge, AMT/K/1/15. Reproduced in (Hodges, 1983, p. 19).

2 This book will refer to the 'imitation game,' the 'Turing test,' and 'Turing's test(s)' interchangeably, as Turing himself did in his communications and writings. It turns out that Turing presented several versions of the imitation test for machine intelligence not only throughout his 1950 text but also before and after it, as will be shown later.

3 A close reading of Turing's exact words in the presentation of his imitation game and test is presented later in Chapter 5.

4 The Cambridge Apostles was an élite and relatively secret intellectual society founded in 1820 (Lubenow, 1999). Many of its members came from King's College, where Turing was a resident and fellow. Notable members include E. M. Forster (1879–1970), Bertrand Russell (1872–1970), and John Maynard Keynes (1883–1946). It is often said that it was a hallmark of the group to discuss philosophical topics in a humorous way.

5 Archives Centre, King's College, Cambridge. AMT/A/5. Turing's testament, signed by him on February 11, 1954.

6 In particular, it is known that in 1933, Turing's final year as an undergraduate in mathematics, Wittgenstein's *Blue Book* and *Brown Book* circulated among mathematics students at Cambridge University (Wittgenstein, 1935, see Preface). Juliet Floyd (2017) opened this perspective and identified important connections between Turing's and Wittgenstein's philosophies in the context of their Cambridge milieu in the early 1930s.

7 See https://academic.oup.com/mind/article/LIX/236/433/986238. Access on 17 June 2023. For comparison, OUP counts 939 citations to the paper 'On Referring' by the eminent philosopher P. F. Strawson (1919–2006), also published in *Mind* in the same year. OUP citation counts for other celebrated *Mind* articles are Charles Darwin's 1877 'A Biographical Sketch of an Infant' (400), William James' 1884 'What is an Emotion?' (2,701), G. E. Moore's 1903 'The Refutation of Idealism' (269), and Bertrand Russell's 1905 'On Denoting' (1,623).

8 This is related to and inspired by a point made by Sterrett (2000).

9 Reference GBR/0272/AMT. Available at: https://archivesearch.lib.cam.ac.uk/repositories/7/resources/1193. Access on 1 June 2023.

10 The Turing Digital Archive is currently available at: http://turingarchive.kings.cam.ac.uk. Access on 1 July 2022.

11 Reference GB 133 NAHC/TUR. See the catalogue at https://archiveshub.jisc.ac.uk/manchesteruniversity/data/gb133-nahc/tur. Access on 1 June 2023.

12 Reference GB 133 TUR/Add. See the catalogue at https://archiveshub.jisc.ac.uk/manchesteruniversity/data/gb133-tur/add. Access on 1 June 2023. The file was an unexpected find for Professor Jim Miles, the history coordinator for the School of Computer Science, who was reorganizing an artifacts storage room

deep in the Kilburn Building in 2017. See http://rylandscollections.com/2017/08/26/remarkable-discovery-of-new-alan-turing-papers/. Access on 1 June 2023.

13 'Alan Turing: The Enigma': https://www.turing.org.uk/. Access on 1 June 2023.

14 A transcript of the notes was later published (1949).

15 In his interview with *The Times*, Turing replies to an argument that not until a machine can write a sonnet 'because of thoughts and emotions felt,' 'could we agree that machine equals brain.' He said that 'the comparison is perhaps a little bit unfair because a sonnet written by a machine will be better appreciated by another machine.'

16 He submitted his paper on morphogenesis to the *Philosophical Transactions of the Royal Society* on November 9, 1951. Published soon after (1952b), he considered this paper 'the equal of "Computable Numbers."'

17 Turing would state in (1948): 'The importance of the universal machine is clear. We do not need to have an infinity of different machines doing different jobs. A single one will suffice. The engineering problem of producing various machines for various jobs is replaced by the office work of "programming" the universal machine to do these jobs' (p. 414).

18 In 1951, Gueroult renamed his chair at the Collège de France 'Histoire et technologie des systèmes philosophiques' ['History and Technology of Philosophical Systems']. Gueroult's works became classics in the history of philosophy in postwar France, influencing thinkers from various schools of thought such as Maurice Merleau-Ponty, Michel Foucault, and Pierre Bourdieu.

19 Turing's early academic mentor Max Newman (1955) wrote: 'Since the paper is easily accessible and highly readable, it would be pointless to summarize it. The conversational style allows the natural clarity of Turing's thought to prevail, and the paper is a masterpiece of clear and vivid exposition' (p. 261). Turing's mother, Mrs. Sara Turing, wrote (1959): 'Almost the whole of this article is within the comprehension of the average reader and is highly entertaining' (p. 94).

20 'It is not a test of intelligence that we need, some critics say, it is a (computational) *theory* of intelligence. This, however, assumes that such a theory is possible—and if intelligence is an emotional concept then such a theory is not possible' (Proudfoot, 2017a, p. 295, her emphasis).

21 In fact, based on the conceptual chronology mentioned above, it can be shown that Turing spent much of his postwar years, 1945–1952, working on a mathematical theory of intelligence.

22 Later, in the 1980s, sociologists of scientific knowledge such as David Bloor (1983) and Harry Collins (1985) also drew inspiration from Wittgenstein's later philosophy, notably from the concepts of a 'language game' and 'forms of life.'

23 For a new point that this book will make, Turing rarely used the word 'artificial,' and when he did, it was to express distaste. For example, he hoped and believed that 'no great efforts' would be put 'into making machines with the most distinctively human, but non-intellectual characteristics, such as the shape of the human body.' He thought of it as 'quite futile' and 'their results would have something like the unpleasant quality of artificial flowers' (Turing, 1951a, p. 486).

24 Recalling his own letter, Malcolm (1958) wrote: 'The article I had read, and of which I wondered whether it was a "leg-pull," was Turing's "Computing Machinery and Intelligence," published in *Mind* 1950. Turing proposed a test for determining whether a machine has successfully simulated human intelligent behaviour. His article gave rise to much subsequent discussion by philosophers of "the Turing test" or "Turing game"' (p. 130, n. 2).

25 In her preface, Cora Diamond (1976) gave the list of attendance of Wittgenstein's 1939 lectures according to the surviving sources: D. A. T. Gasking (1911–1994), Norman Malcolm (1911–1990), R. G. Bosanquet (1918–1944), J. N. Findlay (1903–1987), Casimir Lewy (1919–1991), Marya Lutman-Kokoszynsk (1905–1981), Rush Rhees (1905–1989), Yorick Smythies (1917–1980), Stephen Toulmin (1922–2009), A M. Turing, Alister G. D. Watson, John Wisdom (1904–1993), and Georg Henrik von Wright (1916–2003). There are also references to 'Cunningham' and 'Prince,' whom Diamond was unable to identify. Of all the identified names, except for Turing, all were academic *philosophers*. This includes Alister Watson, then a philosopher of mathematics, who had published a paper in *Mind* a year earlier (Watson, 1938) mentioning 'lengthy discussions' with Turing and Wittgenstein.

26 This analysis of Wittgenstein's quasi-reception of Turing's 1950 paper is indebted to Susan Sterrett and also benefited from Edgar Daylight's feedback, although the responsibility for any errors is mine alone.

27 This archival source and Bertrand Russell's reception of Turing's test, including what he may have meant by 'the general approach' and 'the particular method,' will be revisited and closely examined in Chapter 6.

28 Archives Centre, King's College, Cambridge, AMT/A/29. Jefferson to Turing, March 10, 1951. Reproduced in (Turing, 1959; Hodges, 1983, p. 101; pp. 438–439).

29 Much later (2001), Mays would report that in the summer of 1950 he was asked by Gilbert Ryle to write a reply to Turing's paper to appear in the same October issue of *Mind*. Mays wrote the paper, but according to him Ryle rejected it on the grounds that it was 'too polemical.' So, Mays sent his paper (1952) elsewhere.

30 For example, Turing's focus on morphogenesis led him to decline an invitation from a colleague at Princeton University to contribute to a volume on cybernetics. More on this at the end of this section.

31 Mays, Jefferson, Turing, and others met in a seminar on minds and computing machines at the Philosophy Department of the University of Manchester on October 27, 1949, and possibly again in December 1949, as will be discussed in Chapter 4.

32 'We feel perhaps that we are being pushed, gently not roughly pushed, to accept the great likeness between the actions of electronic machines and those of the nervous system. At the same time, we may misunderstand this invitation and go beyond it to too ready an affirmation that there is identity' (Jefferson, 1949a, p. 1105).

33 'No Mind for Mechanical Man,' *Times*, June 10, 1949, p. 2.

34 The distinction between *think*$_1$ and *think*$_2$ is introduced early in Chapter 1.

35 Archives Centre, King's College, Cambridge, AMT/D/5. Rupert Crawshay-Williams to Turing, April 19, 1951.

36 Instead of citing Wittgenstein directly, Pinsky preferred to cite B. A. Farrell's critique of Wittgenstein (Farrell, 1946).

37 At about the same time that Pinsky's paper was published by Ryle, Ryle himself gave a lecture on the BBC's Third Programme in honor of Wittgenstein, who had died earlier that year. Soon after, the typescript appeared in *Analysis* (1951). At the end, Ryle wrote: 'Wittgenstein has made our generation of philosophers self-conscious about philosophy itself. ... Maybe we have been made a bit neurotic about the nature of our calling' (pp. 8–9).

38 Claude Shannon and John McCarthy to Turing, May 18, 1953. Alan Turing Papers (Additional), University of Manchester Library, GB 133 TUR/Add/123.

39 Turing to Shannon and McCarthy, June 3, 1953. Alan Turing Papers (Additional), University of Manchester Library, GB 133 TUR/Add/123.

40 This term was used in the Ph.D. thesis from which this book is derived, submitted in late 2020 (Gonçalves, 2021, pp. 120–121). It does not refer to the later term 'stochastic parrot,' as used by Bender et al. (2021) to make claims about understanding, meaning, and the use of language that do not seem to seek a dialogue with the philosophy of mind and language.

41 It is worth noting that the term 'artificial intelligence' is post-Turing, being coined later in the United States, although Turing did argue for the possibility of intelligent or thinking machines and thus initiated the discussions on the subject.

42 Herbert Feigl (1960), in his own contribution to the symposium, wrote: 'Any serious effort toward a consistent, coherent, and synoptic account of the place of mind in nature is fraught with embarrassing perplexities. Philosophical temperaments notoriously differ in how they react to these perplexities. Some thinkers apparently like to wallow in them and finally declare the mind-body problem unsolvable: "*Ignoramus et ignorabimus*"' (p. 24).

43 For example, in one of his BBC broadcasts, Turing (1951a) said: 'In this talk ... I shall give most attention to the view which I hold myself, that it is not altogether unreasonable to *describe* digital computers as brains' (p. 482, emphasis added).

44 Chapter 5 argues that the Turing test argument, early in the mid-twentieth century and perhaps from the vantage point of his real-life sensibilities, emphasizes that these kinds of intelligence profiles are *cultural* stereotypes based on dominant but biased views of society and nature.

45 This aspect of the Turing test argument is appears in the discussions of Chapters 3, 4, 5, and 7.

46 Chapter 7 relates this to Turing's social and cultural values, which were certainly different from those of Lucas.

47 Turing opened his paper (1950), writing: 'I propose to consider the question, "Can machines think?" This should begin with definitions of the meaning of the terms 'machine' and 'think' ... Instead of attempting such a definition, I shall replace the question by another, which is closely related to it and is expressed in relatively unambiguous words' (p. 433). Later in (1952), he said: 'I don't want to give a definition of thinking, but if I had to I should probably be unable to

say anything more about it than that it was a sort of buzzing that went on inside my head' (p. 494).

48 Hodges (1983) interpreted Turing's *man-imitates-woman* game as 'a red herring, and one of the few passages of the paper that was not expressed with perfect lucidity' (p. 415).

49 Chapters 5 and 6 will revisit the discussion of such practical Turing tests. The focus here is to appreciate the impact of the Loebner Prize competition in the reception history of the Turing test.

50 'Annual Minsky Loebner Prize Revocation Prize 1995 Announcement,' March 2, 1995. Available at: https://groups.google.com/g/comp.ai/c/dZtU8vDD_bk/m/QYaYB18qAToJ. Access on 10 January 2023.

51 A few years later, Ford and Hayes (1998) insisted in their argument to rethink the goals of AI, claiming that 'the greatest value of artificial intelligence may lie not in imitating human thinking but in extending it into new realms.'

52 On gender specifically, Sterrett would later emphasize (2012; 2017), it can be noted that Turing understood as early as 1950 that gendered behavior is taught and learned in the child's upbringing and would require intelligence to be imitated.

53 '359. Could a machine think? — Could it be in pain? — Well, is the human body to be called such a machine? It surely comes as close as possible to being such a machine. 360. But surely a machine cannot think! — Is that an empirical statement? No. We say only of a human being and what is like one that it thinks. We also say it of dolls; and perhaps even of ghosts. Regard the word "to think" as an instrument!' (Wittgenstein, 1953)

54 AI scientists Hector Levesque (2011) and later Levesque et al. (2012) proposed that approach, inspired by Terry Winograd (1972).

55 See for example Gunderson (1964a). The Turing-Descartes connection has also been made from the broader perspective of the history of philosophy; see for example Copeland (2000a) and Shieber (2004).

56 This is discussed in Chapters 4 and 5.

57 As noted at the end of §1.5.1 from the methodological perspective of the history of philosophy, Proudfoot selects and removes these passages from their context in their exegetical units in order to fit an interpretation, rather than examining their role in the logical development of each unit.

58 Other interpretations can be noted, for instance, James Moor's inductivist interpretation of Turing's proposal (1976). According to Moor, 'the real value of the imitation game lies not in treating it as the basis for an operational definition but in considering it as a potential source of good inductive evidence for the hypothesis that machines think' (p. 249). This was the first of several interpretations that followed, idealizing the Turing test as a practical experiment to determine the existence of machine intelligence.

59 Chapter 4 also contributes to debunking the cartoons by contextualizing Turing's test proposal back into its historical complexity.

60 Chapter 7 gives a broader perspective on Turing's intellectual project.

61 In (1952), he said: 'The important thing is to draw a line between the properties of a brain … that we want to discuss, and those that we don't. To take an

extreme case, we are not interested in the fact that the brain has the consistency of cold porridge' (pp. 494–495).

62 Susan Sterrett (2012, 2017) gives excellent accounts of what I have called experiential and social dimension of learning according to Turing.

63 He seems to have developed this view from his earlier conception and use of heuristics during his wartime service at Bletchley Park. The connection is that learning can be seen as a heuristic process and, as such, errors would be inherent. A heuristic technique is a shortcut for pruning the search space of a mathematical problem. Usually, there is no guarantee that the pruning will not miss out on a solution. Copeland (2017) draws attention to Turing's use of heuristics in the war as a form of machine intelligence, although he does not conceive of it as a form of learning.

64 Norbert Wiener (1894–1964) reported in his famous *Cybernetics* (1948) that he met Turing in the spring of 1947 (p. 23). Wiener made several references to Turing. These include his recognition of Turing's originality: 'Turing, who is perhaps first among those who have studied the logical possibilities of the machine as an intellectual experiment' (p. 13); and his reference to Turing's results from Turing's paper 'On Computable Numbers' (1936) to conclude that 'the logic of the machine resembles human logic, and, following Turing, we may employ it to throw light on human logic' (pp. 125–126). Following the latter passage, Wiener proposed a positive answer to the possibility that the machine might even have 'a more eminently human characteristic,' namely 'the ability to learn.' In doing so, Wiener publicly stated that he shared Turing's non-obvious view that machines could be made to learn for themselves.

65 Christof Teuscher (2002) studied formal properties of Turing's connectionist models.

66 Turing wrote in a letter to B. H. Wood dated November 20, 1951, that he had already sent his 'Chess' manuscript to appear in Vivian Bowden's volume, *Faster than Thought* (1953). 'Turing Additional Papers' GB 133 TUR/Add, item TUR/ADD/43, John Rylands Collections at the University of Manchester Library.

67 Later in 1950, Turing would call them the 'heads in the sand' and the 'theological' objections, respectively.

68 Susan Sterrett (2000) opened this perspective on the Turing test and has since given detailed analyses of what is required of think$_2$ to successfully imitate think$_1$ (Sterrett, 2012, 2017, 2020).

69 Turing to Worsley, June 11, 1951, my emphasis. Unpublished writings of Alan Turing, copyright The Provost and Scholars of King's College Cambridge 2023. B.H. Worsley Collection, 1946–1959, Archives Center, National Museum of American History. Reproduced with permission. I am indebted to Mark Priestley for locating and kindly sharing this source.

70 Chapter 4 details Turing's transition from chess to conversation as the chosen intellectual testbed for machine intelligence.

71 Chapter 4 introduces this source in more detail.

72 This issue was nicely captured by Geoffrey Sampson (1973) in his 'In Defence of Turing' in the 1970s: 'the fact that computer behaviour is known to be determined while *human* behaviour is not can easily be explained: Computers are

designed by humans; humans are not designed by humans' (p. 174, his emphasis). Turing may have been encouraged by the *Blue Book* (Wittgenstein, 1953) to see the human mind and its powers from a naturalistic perspective rather than a dualistic one: 'There is an objection to saying that thinking is some such thing as an activity of the hand. Thinking, one wants to say, is part of our "private experience." It is not material, but an event in private consciousness' (p. 16).

73 Chapter 5 details Turing's use of his test in this aspect of his reply to Jefferson in Section 5.5.1.

74 The so-called Moore's law, the historical observation that the number of transistors in an integrated circuit doubles about every two years, was yet to be stated in the mid-1960s.

75 For a recent survey, for example, of the attempt to mimic the structure and function of artificial neural networks on a silicon integrated circuit, see (Ham et al., 2021).

76 Turing to Worsley, *c.* June, 1951, his emphasis. Unpublished writings of Alan Turing, copyright The Provost and Scholars of King's College Cambridge 2023. B.H. Worsley Collection, 1946–1959, Archives Center, National Museum of American History. Reproduced with permission. Thanks to Mark Priestley for locating and kindly sharing this source.

77 He uses the term 'simulation,' but his use of this term can be taken as synonymous with his technical concept of imitation.

78 This chapter is based on the article by Gonçalves, B. (2022), 'Can machines think? The Controversy that Led to the Turing Test,' *AI & Society* (forthcoming). doi: 10.1007/s00146-021-01318-6). Reproduced here with permission by Springer Nature.

79 At about the same time as my work first appeared (Gonçalves, 2021), Peter Millican, who was an editor of the anthology in which Robin Gandy's testimony appears, also suggested that Gandy's note should have some bearing on how to read Turing's 1950 paper (2021, p. 39).

80 Cf. 'Marvin Minsky on AI: the Turing test is a joke!,' Interview for the Singularity Weblog. Available at: http://www.singularityweblog.com/marvin-minsky/, from 23'35" to 24'45". Access on 1 July 2021.

81 See (Turing, 1950, p. 450); and (Turing, 1951a, pp. 482, 485).

82 Earlier, in February 1946, Hartree had played a key role in securing the Royal Society funding for Newman's Computing Machine Laboratory in Manchester; see (Rope, 2010; Lavington, 2022).

83 '"ACE" will speed jet flying,' *Daily Telegraph*, November 7, 1946.

84 As Chapter 3 notes, Wiener reported (p. 23) to have met Turing in the spring of 1947 and referred to Turing's results from his paper 'On Computable Numbers' (1936) to conclude that 'the logic of the machine resembles human logic, and, following Turing, we may employ it to throw light on human logic' (pp. 125–126). Following this passage, Wiener proposed a positive answer to the possibility that the machine might even have 'a more eminently human characteristic,' namely 'the ability to learn.' Wiener's *Cybernetics* did not go unnoticed in Britain, as Jefferson reacted explicitly respond to it (Section 4.6).

85 See Turing's formulation of problem 10 where he asked the question 'Can the machine play chess?' in his technical report to the National Physical Laboratory (1945, p. 389).

86 At the Polanyi Archive at the University of Chicago, Polanyi scholar Paul Blum found a printed copy of that text containing a few critical annotations by hand, which may indeed have been made by Turing. See (Blum, 2010, p. 52).

87 Polanyi's note: 'in a communication to a Symposium held on "Mind and Machine" at Manchester University in October 1949. This is foreshadowed in "Systems of Logic based on Ordinals," *Proceedings of the London Mathematical Society, Series 2 45*, 1938–1939, pp. 161–228.'

88 '[If] the machine was being put through one of my imitation tests, it would have to do quite a bit of acting...' (Turing et al., 1952, p. 503).

89 'No mind for mechanical man,' *Times*, June 10, 1949, p. 2.

90 Correspondence from Lyn Irvine, wife of Turing's colleague and then director of the University of Manchester Computing Laboratory Max Newman, to Antoinette Esher on June 24, 1949. Reproduced by William Newman (2012).

91 'Calculus to sonnet,' *Times*, June 11, 1949, p. 4.

92 Jules Yule Bogue to Warren McCulloch, *c.* December 1949, Christmas greetings letter found and transcribed by Jonathan Swinton; original in Warren McCulloch archive, MIT American Philosophical Society. Facsimile available at: http://www.manturing.net/manufacturing-blog/2019/6/3/manchester-minds-and-mit-ones. Access on 1 July 2023.

93 Archives Centre, King's College, Cambridge. AMT/B/44.

94 As mentioned in (Section 4.5), Turing made explicit reference to '[his] imitation tests.'

95 This chapter is based on Gonçalves, B. (2023). The Turing test is a thought experiment. *Minds & Machines 33*(1): 1–31. doi: 10.1007/s11023-022-09616-8

96 Turing wrote about the 'imitation game' centrally and extensively throughout his text (1950), but apparently retired the term thereafter. He referred to '[his] test' four times — three times in pp. 446–447 and once on p. 454. He also referred to it as a 'criterion for "thinking"' (p. 436) and as an 'experiment' — once on p. 436, twice on p. 455, and twice again on p. 457 and used the term '*viva voce*' (p. 446). In later sources, Turing referred to a 'viva-voce examination' (1951a, p. 484) and multiple times to '[his] test' (1952), including 'one of my imitation tests' (p. 503). In this chapter, following Turing's own usage, I will refer to all of these terms indistinguishably.

97 Shieber (1994a) made an informative analogy with the Kremer Prize for human-powered flying inspired by the designs of da Vinci. A cash prize is an appropriate incentive in this case because 'the task is just beyond the edge of current technology,' Shieber observed (p. 74). He noted that 'limited tests are better addressed in the near term by engineering (building bigger springs) than science (discovering the airfoil)' (p. 77) and suggested that there still is a substantial scientific gap to be filled in AI.

98 The question of whether Turing, the mathematician, would suggest a subjective criterion for justifying an intelligence claim will be addressed later (Section 5.6.1).

99 Mach is often acknowledged as the thinker who established the use of the term *Gedankenexperiment* ('thought experiment') in the modern scientific tradition.

100 Future work may explore the historical and analytical roots of Turing's familiarity with thought experiments. Juliet Floyd identifies (2017) the intellectual origins of Turing's analysis of computability with the Cambridge tradition of 'common sense,' particularly during Turing's formative years at Cambridge University in the early 1930s. 'Logic was approached,' Floyd writes, 'not first and foremost axiomatically, but practically and in thought experiments' (p. 106).

101 A similar prospect has been suggested by Kuhn (1964) in his analysis of Galileo's thought experiment appearing at the start of 'The First Day' in the *Dialogue Concerning the Two Chief World Systems*: 'For his purpose in this part of the *Dialogue*, it is quite sufficient that we may suppose these things [viz., uniformly accelerated motion and equal instantaneous velocities of the bodies at the bottom of their fall] to be the case' (pp. 251–252). Roy Sorensen (1992) also singled out this as a distinguished property of thought experiments compared to practical experiments. He defined thought experiments as 'experiments that purport to deal with their questions by contemplation of their design rather than by execution' (p. 6). More recently, Michael Stuart (2018) developed a related view of the power of thought experiments in establishing not necessarily new (justified) knowledge but understanding. Stuart leveraged two decades of results in the epistemology of understanding. Further work may build upon the contributions of this chapter to extend the analysis of Turing's imitation tests in connection with the most recent literature on thought experiments in science and philosophy (e.g., Stuart et al., 2018).

102 Mach's specific words in German are: 'Wie man sieht, ist die Grundmethode des Gedankenexperimentes, ebenso wie jene des physischen Experimentes, die Methode der *Variation*. Durch wenn möglich kontinuierliche Variation der Umstände wird das Geltungsbereich einer an dieselben geknüpften Vorstellung (Erwartung) erweitert; durch Modifikation und Spezialisierung der ersteren wird die Vorstellung modifiziert, spezialisiert, bestimmter gestaltet; und diese beiden Prozesse wechseln.' (no emphasis added.)

103 The German words *Vorstellungen* and *Umstände* persisted throughout Mach's original text and were translated to English as 'ideas' and 'conditions'. Alternatively, they could be translated as 'mental images' and 'circumstances.' Note that Mach's seminal text on thought experiments comes as a chapter of his book *Knowledge and Error: Sketches on the Psychology of Enquiry*.

104 Mach neglects that the contradiction can only be empirically verified by assuming the movement takes place in a vacuum, despite that Aristotle's view of motion echoed by Simplicio in the *Two New Sciences* applies to the fall of bodies in media. For a detailed discussion, see John Norton (1996, p. 20). None of this compromises Mach's analysis of thought experiments.

105 Since Mach, there have been substantial accounts of how Galileo connected his law of free fall with his experiments on inclined planes, transferring empirical evidence from the latter to the former. Based on his Galilean studies, Alexandre Koyré states (1953): 'It is well known with what extreme ingenuity, being unable to perform direct measurements, Galileo substitutes for the free fall the motion on an inclined plane on one hand, and that of the pendulum on the

other' (p. 224). After the controversies with Stillman Drake on whether Galileo could have ever validated some of his empirical claims, R. Naylor's study (1974) came to confirm Koyré's point on both historiographical and empirical grounds. This indicates the depth of Mach's insight that variation is the fundamental method of thought experiments and that these 'lie at the basis of science and consciously aim at widening experience' (1897, pp. 135–136).

106 Butler's 'The Book of the Machines' appears in the bibliography of Turing's 1950 paper. It is also known from Turing's mother (1959): 'In his late teens he read [...and] had a particular fondness for The Pickwick Papers, George Borrow's books and Samuel Butler's Erewhon.' This last possibly set him to think about the construction of an actual intelligent machine (p. 108). The Turing-Butler connection is further explored in Chapter 7.

107 Turing's study (1952b) revealed what is called today Turing 'structures' or 'patterns.' Never been observed in nature at the time, they were later experimentally verified and by now have been observed in objects ranging from biological tissues to sand dunes, also appearing at the atomic level (Fuseya et al., 2021).

108 To be precise, Turing's computability theory conceived of 'a fluidity between hardware, software and data' (Floyd, 2017, p. 104), with different elements being explored in different studies.

109 Turing's interest in referring to digital computers in his argument can also be attributed to their popularization by the media in the context of a public controversy on their meaning and significance. Turing, as shown by Floyd (2017), held a strong and overarching interest in connecting advanced technical knowledge with the common language of 'the man in the street.'

110 Sterrett (2000) opened this perspective on Turing's imitation game (cf. Sections 5.5.2 and 5.6.2). As Floyd (2017) noted, Turing was concerned with *types* — 'the delimited, surveyable ordering and organizing of objects, concepts, terms, logical particles, definitions, proofs, procedures and algorithms into surveyable wholes' (p. 104) — and their connections with common sense.

111 Floyd (2017) opened this perspective on Turing's thought, writing: '[h]e saw the difference in levels and types as a complex series of systematizations sensitive to everyday "phraseology" and common sense, not a divide of principle. This was because he always saw "types" or "levels" as lying on an evolving continuum, shaped by practical aspects, the user end, and mathematics' (p. 142). This aspect of Turing's thought, Floyd suggests, can be attributed to the Cambridge tradition of "common sense."'

112 Mach referred to the exclusion of certain conditions as 'mentally diminishing to zero one or several conditions that quantitatively affect the result, so that the remaining factors alone must be taken as of influence' (1897, p. 140). The quantitative result in Turing's case is the interrogator's success rate in making the correct identification of the players.

113 Hartree expressed it in his inaugural lecture at Cambridge University (1947, p. 21) in *c*. November 1946, and later in his *Calculating Instruments and Machines* (1949, p. 70), then attributing it to Ada Lovelace. Turing responded to it first in his lecture to the London Mathematical Society (1947, p. 392) and again in his formulation of 'Lady Lovelace's Objection' (1950, p. 450). The debate would reappear in their radio broadcasts in May 1951 (Jones, 2004).

114   In (1958, p. 261), Polanyi referred to Turing's 1949 argument based on machine chess in the Manchester 'Mind and the Computing Machine' seminar (cf. also Mays, 2000).

115   That passage was quoted in full by Turing (1950) in his discussion of the fourth objection (argument from consciousness), which he attributed to Jefferson (pp. 445–446).

116   The Oxford English Dictionary's definition of 'machine' in the early 1950s (cf. Mays, 1952, p. 149) implies that machine behavior was synonymous with unintelligent behavior, and intelligence was considered an intrinsic property of humankind.

117   As mentioned, Floyd (2017) contextualized Turing's interest in the common use of language.

118   Referring to a preliminary experiment with machine chess, Turing remarked that '[p]laying against such a machine gives a definite feeling that one is pitting one's wits against something alive' (1948, p. 412).

119   Sterrett (2017) draws attention to and explores this aspect of Turing's view of intelligence.

120   For a survey on how AI applications can exploit human-built Internet resources, see (Hovy et al., 2013).

121   Sterrett (2012) presented what appears to be the most substantial account of Turing's views on 'child machines.'

122   'These are possibilities of the near future,' Turing wrote in (1950), 'rather than Utopian dreams' (p. 449).

123   Turing said in (1952), e.g., that the machine 'would be permitted all sorts of tricks so as to appear more man-like' (p. 495), and 'it would have to do quite a bit of acting' (p. 503).

124   Turing's explicit references to 'imitation game,' 'test' and 'experiment' are listed in the first note of Chapter 5.

125   When questioned about the chatbot, which relied on information-retrieval tricks to deceive observers, Warwick and Shah (2015) insisted that their Turing test experiment was designed 'as set out by Alan Turing.'

126   In this case, a one-way crucial experiment, because according to this view while a positive result would be a decisive, confirmatory event for the existence of an intelligent machine, negative results are trivial, since they cannot exclude the possibility of an intelligent machine in the future. This epistemological asymmetry is justified by the tautological fact that 'machines can think' is an open-ended, 'strictly existential statement' (Popper, 1959, pp. 47–48). The proposition cannot be refuted, since no run of the imitation game, or any conceivable experiment, will ever be able to prove empirically that machines cannot think.

127   Shlomo Danziger (2022) explores this point in Turing's thought in detail. I interpret his contribution as strengthening a revealing aspect in Turing's philosophy previously identified by Juliet Floyd (2017), namely the importance of common sense. Here I explore Turing's mild attitude towards the power of a *controlled* experiment, the result of which is, of course, to be interpreted by specialized intellectuals. I emphasize that in considering the possibility of a positive answer to his original question, 'Can machines think?,' Turing shifts the focus of decision-making to 'general educated opinion.' Furthermore, in considering

the possibility of a positive answer to his 'more accurate form of the question,' the imitation game, he proposes that the interrogator and judge should not be an expert, thus making the non-expert profile a structural element of the test's design. This can be seen as one of the elements that makes the Turing test a thought experiment, rather than a controlled experiment (cf. Chapter 5).

128 Simon Lavington (2019) refers to Turing's correspondence from Manchester since April 1949 to emphasize his frustration with the delays in the availability of MADM, the Manchester Automatic Digital Machine, which he helped to design (see, for example, p. 25).

129 When discussing the fifth objection to machine intelligence, the 'Arguments from various disabilities,' Turing wrote (1950): 'Usually if one maintains that a machine can do one of these things, and describes the kind of method that the machine could use, one will not make much of an impression. It is thought that the method (whatever it may be, for it must be mechanical) is really rather base' (pp. 449–450).

130 When asked how a machine could learn by analogy, Turing argued (1952): 'I've certainly left a great deal to the imagination. If I had given a longer explanation I might have made it seem more certain that what I was describing was feasible, but you would probably feel rather uneasy about it all, and you'd probably exclaim impatiently, "Well, yes, I see that a machine could do all that, but I wouldn't call it thinking." As soon as one can see the cause and effect working themselves out in the brain, one regards it as not being thinking, but a sort of unimaginative donkey-work' (p. 500).

131 Moreover, although Turing had practical experience in designing, building, and using high-speed computing machines at Bletchley Park during World War II, and later at the National Physical Laboratory and Manchester University, there seems to be nothing to suggest that he would have ascribed any role to experiment in these cases other than exploring practical and theoretical possibilities. As noted above, he seems to have found 'the actual production of machines' more convincing for demonstration purposes than any particular controlled experiment.

132 Max Newman (1955) wrote: 'Since the paper is easily accessible and highly readable, it would be pointless to summarize it. The conversational style allows the natural clarity of Turing's thought to prevail, and the paper is a masterpiece of clear and vivid exposition' (p. 261). Turing's mother wrote (1959): 'Almost the whole of this article is within the comprehension of the average reader and is highly entertaining' (p. 94).

133 The Oxford English Dictionary's definition of 'machine' in the early 1950s (cf. Mays, 1952, p. 149) implies that machine behavior was synonymous with unintelligent behavior, and intelligence was considered an intrinsic property of humans.

134 Turing wrote: 'Those who hold to the mathematical argument would, I think, mostly be willing to accept the imitation game as a basis for discussion' (p. 445).

135 'I am not very impressed with theological arguments whatever they may be used to support. Such arguments have often been found unsatisfactory in the past. In the time of Galileo ...' (Turing, 1950, pp. 443–444).

136 'To do so would of course meet with great opposition, unless we have advanced greatly in religious toleration from the days of Galileo' (Turing, 1951b, p. 475).

137 Archives Centre, King's College, Cambridge, AMT/D/5. Rupert Crawshay-Williams to Turing, April 19, 1951.

138 This topic appears, for example, in Steven Shapin and Simon Schaffer's *Leviathan and the Air Pump* (1985). Thomas Hobbes is eventually excluded from the Royal Society largely for his anti-experimentalism, which contrasted with Robert Boyle's 'modest, humble, and friendly' experimental philosophy.

139 A mentor of Turing's academic career, Newman was also an FRS (Grattan-Guinness, 2013), as were Douglas Hartree, Geoffrey Jefferson, and Michael Polanyi. As Chapters 4 and 5 have shown, Turing's debate with these three interlocutors in particular led to his formulation of his imitation game and test by making a number of concessions to their arguments and demands.

140 Archives Centre, King's College, Cambridge. AMT/K/54. Alan Turing to Sara Turing, February 11, 1937. Reproduced in (Copeland, 2004, pp. 130–131).

141 For a detailed and scholarly rich account of this, the reader is invited to consult Michael Segre (1989a,b) and Stefano Gattei (2019).

142 (Drake, 1978, p. 415) believes that an inferred reply from Galileo to Vincenzo Renieri's letter of March 13, 1641, must have contained the description of the leaning tower experiment.

143 Interestingly, the same has been claimed about the Turing test: 'The [Turing] tests are circular: they define the qualities they are claiming to be evidence for' (Hayes and Ford, 1995, p. 974). In this connection, here is Turing saying on the radio (1952): 'You might call it a test to see whether the machine thinks, but it would be better to avoid begging the question, and say that the machines that pass are (let's say) 'Grade A' machines (p. 495).

144 In 2015, the experiment was performed for a BBC broadcast from NASA's Space Power Facility in Cleveland, Ohio. Available at: https://www.youtube.com/watch?v=E43-CfukEgs. Access on 9 April 2023.

145 His mentions of tower experiments appear in Drabkin's translation (*ibid.*) on pp. 27, 38, 87, 101, 107, 127. The date of Galileo's *De Motu* is uncertain. Most of it is usually dated to *c.* 1589–1592, when Galileo was a professor at the University of Pisa, but it is known that some parts were written earlier (Drabkin, 1960).

146 Translation from (Cooper, 1935, p. 26).

147 A technical description appears in the 'Apollo 15 Preliminary Science Report'. Available at: http://nssdc.gsfc.nasa.gov/planetary/lunar/apollo_15_feather_drop.html. Access on 11 April 2023.

148 Paul Feyerabend (1970) is known as one of the first scholars to call attention to Galileo's propaganda. Michael Stuart (2021) studied what Feyerabend later called 'the epistemology of drama,' inspired by Galileo's tactics, especially his thought experiments. Segre (1988) presents a portrait of Galileo as a politician.

149 Segre (1980) quotes from the fourth day of the *Discorsi* to illustrate what he takes to be emblematic of Galileo's true views on the role of experiment. After proving that a projectile launched with a certain initial speed will reach a maximal range when thrown at an angle of forty-five degrees, Galileo states: 'The knowledge of a single fact acquired through a discovery of its causes prepares

the mind to understand and ascertain other facts without need of recourse to experiment' (cf. *ibid.*, p. 248).

150 Turing implies in this 1950 passage, in continuity with his arguments about the feasibility of exploring machine intelligence in various intellectual fields (1948, pp. 420–421), that chess was bound to yield faster results compared to more complex fields such as the 'learning of languages,' which, while more impressive, would require sophisticated sensory organs to enable the contact with human agents.

151 *Minimum weight design: Memories of Alan Turing.* Dr. R. K. Livesley. Archives Centre, King's College, Cambridge, AMT/C/33.

152 As mentioned above, this is consistent with the interpretation proposed by Danziger (2022).

153 It is known from Donald Bayley, who worked with Turing from 1943 to 1946, that Turing had spoken at the end of the war of his intention 'to build a brain' (Hodges, 1983; Sykes, 1992, p. 290; 25–27'). He joined the National Physical Laboratory in October 1945 to pursue this project, as he suggested in a letter to the cybernetician Ross Ashby (1946).

154 In (1948), Turing wrote: 'The whole thinking process is still rather mysterious to us, but I believe that the attempt to make a thinking machine will help us greatly in finding out how we think ourselves' (p. 486).

155 As mentioned in Chapter 4, Peter Millican, an editor of the anthology in which Robin Gandy's anecdote appears, also thinks that what Gandy reports should have some bearing on how to read Turing's 1950 paper (2021, p. 39).

156 *The Guardian*, June 9, 2014. Scientists dispute whether computer 'Eugene Goostman' passed Turing test. Available at: http://www.theguardian.com/technology/2014/jun/09/scientists-disagree-over-whether-turing-test-has-been-passed. Access on 10 January 2023.

157 The Singularity Blog, July 2013. Marvin Minsky on AI: The Turing Test is a Joke! Available at: http://www.singularityweblog.com/marvin-minsky/, from 23'35" to 24'45". Access on 9 February 2023.

158 With the exception of Jon Agar, who noted Turing's blurring of 'the rigid separation between order givers and order followers,' his 'transgression of hierarchical divisions,' leading to his recasting of the computer 'as a direct threat to generalist "masters"' of the postwar government machine (2003, pp. 267–268).

159 Geoffrey Jefferson to Mrs. Turing, December 18, 1954. Archives Centre, King's College, Cambridge, AMT/A/16. Reproduced in (Turing, 1959, p. 58).

160 As Chapter 2 shows (Section 2.1), even on the occasion of Turing's election to the Royal Society in March 1951, Jefferson could not resist referring to Turing's alleged behaviorism.

161 For example, the spirit of the times was thus expressed by Raymond Aron (1955): 'Every action, in the middle of the twentieth century, presupposes and involves the adoption of an attitude with regard to the Soviet enterprise' (p. 55).

162 In the 1950s, being considered a 'non-conformist' could lead to political and moral charges in the US and Great Britain. For example, in September 1952, Charles Chaplin, returning from a European tour, was banned from entering

the US for political and moral offenses (Hickey, 1969). Later in his autobiography, Chaplin wrote (1964): 'My prodigious sin was, and still is, being a non-conformist.'

163 Turing to Norman Routledge, *c.* mid-February 1952. Archives Centre, King's College, Cambridge, AMT/D/14a. Reproduced in (Turing, 1952a).

164 1983 is the year of the potentially disastrous incident known as the Able Archer 'War Scare.'

165 For a biography of Carpenter, see Sheila Rowbotham (2009).

166 Fearing the impact of Turing-inspired automation on the current economy, Brynjolfsson points to Jeff Bezos' Amazon business model, based on subcontracting and GIGs, as a paradigmatic example of what he calls human augmentation, and urges scientists and policymakers to pursue it.

167 See https://www.imdb.com/title/tt0115749/. Available at: https://www.youtube.com/watch?v=udW0j96vAOk. Access on 15 June 2023.

168 Three years later, in the United States, John McCarthy et al. (1955) did just that in their coining of the term 'artificial intelligence.' As Agar (2012) notes (p. 382), McCarthy sought philanthropic and Cold War patronage for his AI seminar proposal.

169 Lyn Irvine (1901–1973) was a literary critic who kept her family name after marrying mathematics professor Max Newman, who would become Turing's close colleague and career mentor. Irvine was fond, reports her son William Newman (2002), of Turing's 'very simple, humble, gentle personality' (p. 53).

170 For a vivid collection of testimonies and anecdotes by close friends and colleagues of Turing, see Christopher Sykes' BBC documentary (1992).

171 'The Mechanical Brain,' *The Times* (1949, June 14), London, p. 5. Quoted in (Hodges, 1983, p. 406).

172 Note that in objection (2), but not in objection (1), Turing uses quotes to mark the statement of the objection, thereby distinguishing it from his own exposition that follows.

173 Agar also referred to a related nineteenth-century cultural critic, the English poet Matthew Arnold (1869), who urged 'a stream of fresh and free thought upon our stock notions and habits, which we now follow staunchly but mechanically, vainly imagining that there is virtue in following them staunchly which makes up for the mischief of following them mechanically' (p. viii).

174 For an interesting survey of this, see Jack Copeland (2000a).

175 Turing also stated in (1948): 'The importance of the universal machine is clear. We do not need to have an infinity of different machines doing different jobs. A single one will suffice. The engineering problem of producing various machines for various jobs is replaced by the office work of "programming" the universal machine to do these jobs' (p. 414).

176 This is how Turing's contemporary George Orwell saw it in his commentary in the '8 June 1945 Broadcast for Schools: *Erewhon*,' BBC Home Service.

177 This is a core theme of Butler's works *Erewhon* (1872), *Life and Habit* (1878), *Evolution, Old and New; Or, the Theories of Buffon, Dr. Erasmus Darwin, and Lamarck, as compared with that of Charles Darwin* (1879), *Unconscious Memory* (1880), *Luck or Cunning as the Main Means of Organic Modification?* (1887).

178 The chapters II and III of this book were meant to explain how Butler came to write the 'Book of the Machines' in *Erewhon* (Butler, 1880, see the 1910 Introduction by Marcus Hertog).

179 Peter T. Saunders makes a related observation in his introduction to the third volume of the *Collected Works* (1992). Turing's experiences with digital computers, Saunders noted, could 'have led him to see the development of an organism from egg to adult as being programmed in the genes. This would have been in the spirit of the times …' (p. xi). Saunders further noted that the neo-Darwinian 'synthetic theory of evolution had only been completed about ten years earlier' (by Julian Huxley in 1942), 'combining Darwinian natural selection with Mendelian genetics;' he also cited (John von Neumann's 1948) cellular automata, 'models in which the fate of a cell is determined by the states of its neighbors through some simple algorithm, in a way that is reminiscent of the Turing machine.'

180 The next subsection will show that the same may be true of Turing's views on society and institutions. Recall Agar's observation of Turing's blurring of 'the rigid separation between order givers and order followers' and his 'transgression of hierarchical divisions' (2003, pp. 267-268).

181 'Electronic Brains Can't Reason,' *Daily Mail* (1946, November 19), p. 2.

182 This is confirmed by the NPL director, Charles Darwin, grandson of the evolutionary theorist, in his letter to *The Times* dated November 11, 1946, which appeared two days later on p. 7.

183 The ENIAC was built to speed and scale ballistic calculations in World War II and later reused for the simulation of atomic explosions (Fitzpatrick, 1999). Early digital computing projects in Britain were also motivated by scientific and military applications, notably in the context of the British atomic energy project (Lavington, 2019).

184 Charles Darwin to *Times* (op. cit.). Reproduced in (Hodges, 1983, p. 357).

185 *Times*, 'The "Electronic Brain": A Misleading Term; No Substitute for Thought,' November 7, 1946, p. 5.

186 Chapter 4 covers the Turing-Hartere exchanges in more detail.

187 Turing would revisit the issue, first, in his formulation of the Lady Lovelace objection (1950, p. 450), addressing Hartree's mobilization of Lovelace's words delivered over a hundred years ago (1949, p. 70); and again, in one of his two BBC broadcasts (1951a), rehearsing his Dickensian satire of mechanized computing as a military operation, and insisting on a more open interpretation of computers: 'The less doubt there is about what is going to happen [in computing] the better the mathematician is pleased. It is like planning a military operation. Under these circumstances … I agree with Lady Lovelace's dictum as far as it goes, but I believe that its validity depends on considering how digital computers *are* used rather than how they *could* be used' (p. 482, his emphasis).

188 The sex division of labor in the postwar British computer industry is documented by Mar Hicks (2017).

189 Turing's fears would soon come true as the human computers, mostly women and a crucial workforce at Bletchley Park (Hicks, 2017), would be massively replaced by digital computers.

Bibliography

Abramson, D. (2008). Turing's responses to two objections. *Minds and Machines 18*(2), 147–167. doi: 10.1007/s11023-008-9094-6.

Abramson, D. (2011). Descartes' influence on Turing. *Studies in History and Philosophy of Science: Part A 42*(4), 544–551. doi:10.1016/j.shpsa.2011.09.004.

Adler, C. G. and B. L. Coulter (1978). Galileo and the Tower of Pisa experiment. *American Journal of Physics 46*(3), 199–201. doi:10.1119/1.11165.

Agar, J. (2003). *The Government Machine: A Revolutionary History of the Computer*. Cambridge, MA: MIT Press.

Agar, J. (2012). *Science in the Twentieth Century and Beyond*. Cambridge: Polity.

Airey, J. L. (2019). *Religion Around Mary Shelley*. University Park, PA: Penn State University Press.

Anderson, W. (2022). History and philosophy of science takes form. *Studies in the History and Philosophy of Science 93*, 175–182. doi:10.1016/j.shpsa.2022.04.001.

Anscombe, G. E. M. (1965 [1959]). *An Introduction to Wittgenstein's* Tractatus. New York: Harper & Row.

Arnold, M. (1889 [1869]). *Culture and Anarchy: An Essay in Political and Social Criticism* (Popular ed.). London: Elder Smith. Available at: https://archive.org/details/cultureanarchyes00arno. Access on 1 May 2023.

Aron, R. (1962 [1955]). *The Opium of the Intellectuals*. New York: W. W. Norton. Trans. by Terence Kilmartin from the French Edition (1955, Paris: Calmann-Lévy).

Beck, L. W. (1969). Introduction and bibliography. *The Monist 53*(4), 523–531. Available at: http://www.jstor.org/stable/27902145.

Bedini, S. A. (1964). The role of automata in the history of technology. *Technology and Culture 5*(1), 24–42. doi:10.2307/3101120.

Bender, E., T. Gebru, A. McMillan-Major, and M. Mitchell (2021). On the dangers of stochastic parrots: can language models be too big? In *Proceedings of the 2021 ACM Conference on Fairness, Accountability, and Transparency (FAccT'21)*, pp. 610–623. New York: ACM. doi:10.1145/3442188.3445922.

Block, N. (1981). Psychologism and behaviorism. *The Philosophical Review XC*(1), 5–43. doi: 10.2307/2184371.

Bloor, D. (1983). *Wittgenstein: A Social Theory of Knowledge*. Theoretical Traditions in the Social Sciences. London: MacMillan.

Bloor, D. (1991 [1976]). *Knowledge and Social Imagery* (Second ed.). Chicago: University of Chicago Press.

Blum, P. R. (2010). Michael Polanyi: can the mind be represented by a machine? Documents of the discussion in 1949. *Polanyiana 19*(1–2), 35–60. Available at: https://philpapers.org/rec/BLUMPC-2. Access on 3 July 2022.

BMJ (1949). Mind, machine, and man. *British Medical Journal 1*(4616), 1129–1130.

Boden, M. (2006). *Mind as Machine: A History of Cognitive Science*, Volume 1 & 2. Oxford: Oxford University Press.

Bowden, B. V. (Ed.) (1953). *Faster than Thought: A Symposium on Digital Computing Machines.* London: Pitman. Facsimile available at: http://archive.org/details/fasterthanthough00bvbo.

Brewster, E. T. (1912). *Natural Wonders Every Child Should Know*. New York: Doubleday, Doran & Co., Inc.

Bringsjord, S., P. Bello, and D. Ferrucci (2001). Creativity, the Turing Test, and the (better) Lovelace Test. *Minds and Machines 11*, 3–27. doi:10.1023/A:1011206622741.

Brynjolfsson, E. (2022). The Turing trap: The promise & peril of human-like artificial intelligence. *Daedalus 151*(2), 272–287. doi:10.1162/DAED_a_01915.

Buchdahl, G. (1962). History and philosophy of science at Cambridge. *History of Science 1*(1), 62–66. doi:10.1177/007327536200100107.

Bullynck, M., E. G. Daylight, and L. De Mol (2015). Why did computer science make a hero out of Turing? *Communications of the ACM 58*(3), 37–39. doi:10.1145/2658985.

Butler, S. (1872). *Erewhon or Over the Range*. London: Trubner & Co.

Butler, S. (2004 [1880]). *Unconscious Memory*. Project Gutenberg. Transcribed from the 1910 A. C. Fifield edition, London, by David Price. Available at: https://www.gutenberg.org/ebooks/6605. Access on 18 November 2022.

Capek, K. (2019 [1923]). *R.U.R. (Rossum's Universal Robots)*. Toronto: Samuel French. Available at: https://gutenberg.org/ebooks/59112.

Carpenter, B. E. and R. W. Doran (1977). The other Turing machine. *The Computer Journal 20*(3), 269–279. doi:10.1093/comjnl/20.3.269.

Castañeda, H.-N. (1974). Leibniz's concepts and their coincidence salva veritate. *Noûs 8*(4), 381–398.

Chaplin, C. (1966 [1964]). *My Autobiography* (Pocket Books ed.). New York: Simon & Schuster.

Chomsky, N. (1959). Review of B. F. Skinner's *Verbal Behavior*. *Language 35*(1), 26–58. doi:10.2307/411334.

Chomsky, N. (1995). Language and nature. *Mind 104*(413), 1–61. doi:10.1093/mind/104.413.1.

Chomsky, N. (2004). Turing on the 'imitation game'. In S. M. Shieber (Ed.), *The Turing Test: Verbal Behavior as the Hallmark of Intelligence*, Chapter 20, pp. 317–321. Cambridge, MA: The MIT Press.

Clarke, A. C. (1968). *2001: A Space Odyssey*. New York: Signet.

Collins, H. (1985). *Changing Order: Replication and Induction in Scientific Practice*. London: Sage Publications.

Cooper, L. (1935). *Aristotle, Galileo, and the Tower of Pisa*. New York: Ithaca.

Cooper, S. B. and J. van Leeuwen (Eds.) (2013). *Alan Turing: His Work and Impact*. Amsterdam: Elsevier Science.

Copeland, B. J. (1999). A lecture and two radio broadcasts on machine intelligence by Alan Turing. In K. Furukawa, D. Michie, and S. Muggleton (Eds.), *Machine Intelligence 15*. Oxford: Oxford University Press.

Copeland, B. J. (2000a). Narrow versus wide mechanism: including a re-examination of Turing's views on the mind-machine issue. *The Journal of Philosophy* 97(1), 5–32. doi:10.2307/2678472.

Copeland, B. J. (2000b). The Turing test. *Minds and Machines* 10(4), 519–539. doi:10.1023/A:1011285919106.

Copeland, B. J. (Ed.) (2004). *The Essential Turing: The Ideas that Gave Birth to the Computer Age*. Oxford: Oxford University Press.

Copeland, B. J. (2006). Colossus and the rise of the modern computer. In B. J. Copeland (Ed.), *Colossus: The Secrets of Bletchley Park's Codebreaking Computers*, Chapter 9, pp. 101–115. Oxford: Oxford University Press.

Copeland, B. J. (Ed.) (2012a). *Alan Turing's Electronic Brain: The Struggle to Build the ACE, the World's Fastest Computer*. Oxford: Oxford University Press.

Copeland, B. J. (2012b). *Turing: Pioneer of the Information Age*. Oxford: Oxford University Press.

Copeland, B. J. (2017). Intelligent machinery. In B. J. Copeland et al. (Eds.), *The Turing Guide*, Chapter 25, pp. 265–275. Oxford: Oxford University Press.

Copeland, B. J., J. P. Bowen, M. Sprevak, and R. J. Wilson, et al. (2017). *The Turing Guide*. Oxford: Oxford University Press.

Copeland, J. and D. Proudfoot (2009). Turing's test: a philosophical and historical guide. In R. Epstein, G. Roberts, and G. Beber (Eds.), *Parsing the Turing Test: Philosophical and Methodological Issues in the Quest for the Thinking Computer*. Dordrecht: Springer.

Crawshay-Williams, R. (1970). *Russell Remembered*. Oxford: Oxford University Press.

Crease, R. P. (2003). The legend of the leaning tower. *Physics World* (4 Feb.). Available at: http://physicsworld.com/a/the-legend-of-the-leaning-tower/. Access on 6 October 2020.

Danziger, S. (2022). Intelligence as a social concept: a socio-technological interpretation of the Turing test. *Philosophy & Technology* 35(68). doi:10.1007/s13347-022-00561-z.

Darwin, C. (1947, 23 Jul.). Letter from Darwin to Edward Appleton. Technical report, National Physical Laboratory, Teddington. Facsimile available at: http://www.alanturing.net/darwin_appleton_23jul47/. Access on 1 July 2023.

Darwin, C. G. (1958). Douglas Rayner Hartree, 1897–1958. *Biographical Memoirs of Fellows of the Royal Society* 4(Nov.), 102–116. doi: 10.1098/rsbm.1958.0010.

Daylight, E. (2023). True Turing: A bird's eye view. *Minds and Machines*. doi:10.1007/s11023-023-09634-0.

Daylight, E. G. (2014). A Turing tale. *Communications of the ACM* 57(10), 36–38. doi:10.1145/2629499.

Daylight, E. G. (2015). Towards a historical notion of 'Turing – the father of computer science'. *History and Philosophy of Logic* 36(3), 205–228. doi:10.1080/01445340.2015.1082050.

Dennett, D. (2006 [1984]). Can machines think? In C. Teuscher (Ed.), *Alan Turing: Life and Legacy of a Great Thinker*, pp. 295–316. Berlin: Springer. Reprinted from (Ed.) M. G. Shafto, *How We Know*, (San Francisco: Harper &

Row) 121–145, 1984, plus postscripts "Eyes, ears, hands and history" (1985) and (1997, no title).

Dennett, D. (2006 [1997]). Postscript (1997, no title) to "Can machines think?" (1984). In C. Teuscher (Ed.), *Alan Turing: Life and Legacy of a Great Thinker*, pp. 314–316. Berlin: Springer.

Descartes, R. (1985 [1637]). Discourse and essays. In J. Cottingham, R. Stoothoff, and D. Murdoch (Eds.), *The Philosophical Writings of Descartes*, Volume I, pp. 111–151. Cambridge: Cambridge University Press.

Diamond, C. (Ed.) (1976). *Wittgenstein's Lectures on the Foundations of Mathematics, Cambridge, 1939*. Ithaca: Cornell University Press.

Dickens, C. (1857). *Little Dorrit*. London: Bradbury & Evans. Available at: https://archive.org/details/littledorrit00dickrich.

Dickens, C. (1972 [1836]). *The Pickwick Papers*. London: Penguin Classics.

Dijkstra, E. (1984, November). The threats to computing science. Talk delivered at the ACM 1984 South Central Regional Conference, November 16–18, Austin, Texas. Available at: http://www.cs.utexas.edu/users/EWD/transcriptions/EWD08xx/EWD898.html. Access on 20 June 2020.

Drabkin, I. E. (1960). A note on Galileo's *De Motu*. *Isis 51*(3), 271–277. doi:10.1086/348910.

Drake, S. (1973). Galileo's experimental confirmation of horizontal inertia: unpublished manuscripts. *Isis 64*(3), 291–305. doi:10.1086/351124.

Drake, S. (1978). *Galileo at Work: His Scientific Biography*. Chicago: The University of Chicago Press.

Epstein, R. (1992). The quest for the thinking computer. *AI Magazine 13*(2), 81–95. doi: 10.1609/aimag.v13i2.993.

Epstein, R., G. Roberts, and G. Beber (Eds.) (2009). *Parsing the Turing Test: Philosophical and Methodological Issues in the Quest for the Thinking Computer*. Dordrecht: Springer.

Evans, C. R. and A. D. J. Robertson (Eds.) (1968). *Key Papers: Cybernetics*. Baltimore: University Park Press.

Farrell, B. A. (1946). An appraisal of therapeutic positivism. (I.). *Mind 55*(217), 25–48. doi:10.1093/mind/LV.219.25.

Feigl, H. (1960). Mind-body, *not* a pseudoproblem. In S. Hook (Ed.), *Dimensions of Mind: A Symposium*. New York: New York University Press.

Feyerabend, P. (2010 [1970]). *Against Method* (Fourth ed.). London: Verso.

Fitzpatrick, A. (1999). *Igniting the Light Elements: The Los Alamos Thermonuclear Weapon Project, 1942–1952*. Ph. D. thesis, United States. doi:10.2172/10596.

Floridi, L. (2023). AI as *Agency Without Intelligence*: on ChatGPT, large language models, and other generative models. *Philosophy & Technology 36*(1). doi:10.1007/s13347-023-00621-y.

Floridi, L., M. Taddeo, and M. Turilli (2009). Turing's imitation game: Still an impossible challenge for all machines and some judges: An evaluation of the 2008 Loebner Contest. *Minds and Machines 19*(1), 145–150. doi: 10.1007/s11023-008-9130-6.

Floyd, J. (2017). Turing on "common sense": Cambridge resonances. In J. Floyd and A. Bokulich (Eds.), *Philosophical Explorations of the legacy of Alan Turing*, Volume 324 of *Boston Studies in the Philosophy and History of Science*, Chapter 5, pp. 103–149. Cham: Springer. doi:10.1007/978-3-319-53280-6_5.

Ford, K. and P. Hayes (1998). On computational wings: rethinking the goals of artificial intelligence. *Scientific American Presents 9*(4), 78–83.

Franklin, A. (1979). Galileo and the leaning tower: an Aristotelian interpretation. *Physics Education 14*(1), 60–63. doi: 10.1088/0031-9120/14/1/316.

French, R. (1990). Subcognition and the limits of the Turing test. *Mind 99*(393), 53–65. doi: 10.1093/mind/XCIX.393.53.

Fryer, D. M. and J. C. Marshall (1979). The motives of Jacques de Vaucanson. *Technology and Culture 20*(2), 257–269. doi:10.2307/3103866.

Fulton, J. F. (1960). The honourable Robert Boyle, F. R. S. (1627–1692). *Notes and Records of the Royal Society 15*(1). doi: 10.1098/rsnr.1960.0012.

Furbank, N. (1948 [1946]). *Samuel Butler (1835–1902)*. Cambridge: Cambridge University Press. Amended version of Furbank's 1946 Le Bas Prize winning essay.

Fuseya, Y., H. Katsuno, K. Behnia, et al. (2021). Nanoscale Turing patterns in a bismuth monolayer. *Nature Physics 17*, 1031–1036. doi:10.1038/s41567-021-01288-y.

Galilei, G. (1890–1909). *Le opere di Galileo Galilei* (Edizione nazionale ed.), Volume 20. Florence: Barbèra.

Galilei, G. (1960 [c. 1590]). De motu (c. 1590) In I. E. Drabkin and S. Drake (Eds.), *On Motion and On Mechanics*. Madison: University of Wisconsin Press.

Galilei, G. (1974 [1638]). *Two New Sciences*. Madison: University of Wisconsin Press. Translated by Stillman Drake.

Gandy, R. (1996). Human versus mechanical intelligence. In P. Millican and A. Clark (Eds.), *Machines and Thought: The Legacy of Alan Turing*, Volume 1, pp. 125–136. Oxford: Oxford University Press.

Gattei, S. (2019). *On the Life of Galileo: Viviani's Historical Account and Other Early Biographies*. Princeton: Princeton University Press.

Gendler, T. S. (1998). Galileo and the indispensability of scientific thought experiment. *British Journal for the Philosophy of Science 49*(3), 397–424. doi: 10.1093/bjps/49.3.397.

Genova, J. (1994). Turing's sexual guessing game. *Social Epistemology 8*(4), 13–26. doi:10.1080/02691729408578758.

Gödel, K. (1965 [1931]). On formally undecidable propositions of Principia Mathematica and related Systems I. In M. Davis (Ed.), *The Undecidable: Basic Papers on Undecidable Propositions, Unsolvable Problems and Computable Functions*, pp. 5–38. New York: Raven. English translation of K. Gödel, "Über formal unentscheidbare Sätze der Principia Mathematica und verwandter Systeme I", *Monatshefte für Mathematik und Physik* 38: 173–198, 1931.

Gonçalves, B. (2021, March). *Machines Will Think: Structure and Interpretation of Alan Turing's Imitation Game*. Ph. D. thesis, Faculty of Philosophy, Languages and Human Sciences, University of São Paulo, São Paulo. doi:10.11606/T.8.2021.tde-10062021-173217.

Gonçalves, B. (2022). Can machines think? The controversy that led to the Turing test. *AI & Society*. doi:10.1007/s00146-021-01318-6.

Gonçalves, B. (2023a). Galilean resonances: the role of experiment in Turing's construction of machine intelligence. *Annals of Science*. doi:10.1080/00033790.2023.2234912.

Gonçalves, B. (2023b). Irony with a point: Alan Turing and his intelligent machine utopia. *Philosophy & Technology 36*(3). doi:10.1007/s13347-023-00650-7.

Gonçalves, B. (2023c). The Turing test is a thought experiment. *Minds & Machines 33*(1), 1–31. doi:10.1007/s11023-022-09616-8.

Grattan-Guinness, I. (2013). The mentor of Alan Turing: Max Newman (1897–1984) as a logician. *The Mathematical Intelligencer 35*(3), 54–63. doi:10.1007/S00283-013-9387-3.

Gueroult, M. (1984 [1952]). *Descartes' Philosophy Interpreted According to the Order of Reasons*. Minneapolis: University of Minnesota Press.

Gunderson, K. (1964a). Descartes, La Mettrie, language, and machines. *Philosophy 39*(149), 193–222. doi:10.1017/S0031819100005559555595.

Gunderson, K. (1964b). The imitation game. *Mind 73*(290), 234–245. doi:10.1093/mind/LXXIII.290.234.

Ham, D., H. Park, S. Hwang, and K. Kim (2021). Neuromorphic electronics based on copying and pasting the brain. *Nature Electronics 4*, 635–644. doi:10.1038/s41928-021-00646-1.

Hartree, D. R. (1947). *Calculating Machines: Recent and Prospective Developments and their Impact on Mathematical Physics*. Cambridge: University Press. Inaugural lecture in *c.* November 1946.

Hartree, D. R. (1949). *Calculating Instruments and Machines*. Urbana: University of Illinois Press.

Haugeland, J. (1979). Understanding natural language. *Journal of Philosophy 76*(11), 619–632. doi: 10.2307/2025695.

Hayes, P. and K. Ford (1995). Turing test considered harmful. In *Proceedings of the 14th International Joint Conference on Artificial Intelligence (IJCAI'95)*, pp. 972–977.

Hickey, T. (1969). Accusations against Charles Chaplin for political and moral offenses. *Film Comment 5*(4), 44–57. JSTOR:43754277.

Hicks, M. (2017). *Programmed Inequality: How Britain Discarded Women Technologists and Lost Its Edge in Computing*. Cambridge, MA: MIT Press.

Hoare, T. (2003). Edsger Wybe Dijkstra. *Physics Today 56*(3), 96. doi: 10.1063/1.1570789.

Hodges, A. (1983). *Alan Turing: The Enigma*. London: Simon & Schuster.

Hodges, A. (2008). In retrospect: Gödel's proof. *Nature 454*(829). doi:10.1038/454829a.

Hodges, A. (2009). Alan Turing and the Turing test. In R. Epstein, G. Roberts, and G. Beber (Eds.), *Parsing the Turing Test: Philosophical and Methodological Issues in the Quest for the Thinking Computer*, Chapter 2, pp. 13–22. Berlin: Springer.

Hook, S. (Ed.) (1960). *Dimensions of Mind: A Symposium*. New York: New York University Press. Available at: https://archive.org/details/dimension sofmind013653mbp. Accessed 1 June 2023.

Hovy, E., R. Navigli, and S. P. Ponzetto (2013). Collaboratively built semi-structured content and Artificial Intelligence: the story so far. *Artificial Intelligence 194*, 2–27. doi:10.1016/j.artint.2012.10.002.

Irvine, L. (1949). Lyn Newman to Antoinette Esher, June 24, 1949. Personal Papers of Lyn Newman, Girton College Archive, Cambridge.

Irvine, L. (2012 [1959]). Foreword to the first edition. In *Alan M. Turing: Centenary Edition*, pp. xix–xxiv. Cambridge: Cambridge University Press.

Jefferson, G. (1949a). The mind of mechanical man. *British Medical Journal 1*(4616), 1105–1110. doi:10.1136/bmj.1.4616.1105.

Jefferson, G. (1949b, September). René Descartes on the localisation of the soul. *Irish Journal of Medical Science* (285), 691–706.

Jones, A. (2004). Five 1951 BBC broadcasts on automatic calculating machines. *IEEE Annals of the History of Computing 26*(2), 3–15. doi:10.1109/MAHC.2004.1299654.

Koyré, A. (1937). Galilée et l'expérience de Pise. In *Annales de l'Université de Paris*, Volume 12. BnF. Available at: http://gallica.bnf.fr/ark:/12148/bpt6k938880.

Koyré, A. (1953). An experiment in measurement. *Proceedings of the American Philosophical Society 97*(2), 222–237. JSTOR:3143896.

Koyré, A. (1977 [1939]). *Galileo Studies*. New Jersey: The Harvester Press. Translated by J. Mepham from *Études galiléennes*, Paris: Hermann, 1939.

Kuhn, T. (1977 [1964]). A function for thought experiments. In T. Kuhn (Ed.), *The Essential Tension: Selected Studies in Scientific Tradition and Change*, pp. 240–265. University of Chicago Press. Reprinted from (Ed.) Alexandre Koyré, *Mélanges Alexandre Koyré, publiés à l'occasion de son soixante-dixième anniversaire: L'aventure de la science*, (Paris: Hermann, 1964), 2:307–334.

Lavington, S. (2012). *Alan Turing and His Contemporaries: Building the World's First Computers*. Swindon: British Conservation Society: the Chartered Institute for IT.

Lavington, S. (2019). *Early computing in Britain: Ferranti Ltd and government funding, 1948–1958*. History of Computing. Cham: Springer.

Lavington, S. (2022). Early days of computing at manchester: Max Newman's Royal Society project, 1946–1951. *IEEE Annals of the History of Computing 44*(2), 20–32. doi:10.1109/MAHC.2022.3159738.

Levesque, H. (2011). The Winograd schema challenge. Commonsense 2011. Available at: http://commonsensereasoning.org/2011/papers/Levesque.pdf. Access on 1 July 2023.

Levesque, H., E. Davis, and L. Morgenstern (2012). The Winograd schema challenge. In *Proceedings of the Thirteenth International Conference on Principles Of Knowledge Representation And Reasoning (KR'2012)*, pp. 552–561. AAAI.

Longo, G. (2018). Letter to Turing. *Theory, Culture & Society 36*(6), 73–94. doi:10.1177/0263276418769733.

Lubenow, W. C. (1999). *The Cambridge Apostles 1820–1914: Liberalism, Imagination, and Friendship in British Intellectual and Professional Life*. Cambridge: Cambridge University Press.

Lucas, J. R. (1961). Minds, machines and Gödel. *Philosophy 36*(137), 112–127. doi:10.1017/S0031819100057983.

Mach, E. (1976 [1897]). On thought experiments. In E. N. Hiebert (Ed.), *Knowledge and Error: Sketches on the Psychology of Enquiry*, Vienna circle collection, Chapter 11, pp. 134–147. Dordrecht-Holland: D. Reidel. Translation of the 5th edition of *Erkenntnis und Irrturn* (Leipzig: Johann Ambrosius Barth, 1905), which included "Über Gedankenexperimente", in: *Zeitschrift für den physikalischen und chemischen Unterricht*, 10: 1–5., 1897. doi: 10.1007/978-94-010-1428-1.

MacLachlan, J. (1973). A test of an 'imaginary' experiment of Galileo's. *Isis 64*(3), 374–379. doi:10.1086/351130.

Malcolm, N. (2001 [1958]). *Ludwig Wittgenstein: A Memoir* (Second ed.). Oxford: Clarendon Press.

Manuel, F. E. and F. P. Manuel (1979). *Utopian Thought in the Western World.* Cambridge, MA: Harvard University Press.

Marcus, G., F. Rossi, and M. Veloso (2016). Beyond the Turing test. *AI Magazine 37*(1), 3–4. Special issue editorial. doi: 10.1609/aimag.v37i1.2650.

Mays, W. (1952). Can machines think? *Philosophy 27*(101), 148–162. doi: 10.1017/S003181910002266X.

Mays, W. (1967). Recollections of Wittgenstein. In K. T. Fann (Ed.), *Ludwig Wittgenstein: The Man and His Philosophy.* Sussex: Humanities Press.

Mays, W. (2000). Turing and Polanyi on minds and machines. *Appraisal 3*(2), 55–62.

Mays, W. (2001). My reply to Turing: fiftieth anniversary. *Journal of the British Society for Phenomenology 32*(1), 4–23. doi:10.1080/00071773. 2001.11007314.

McCarthy, J. (2007). From here to human-level AI. *Artificial Intelligence 171*(18), 1174–1182. doi:10.1016/j.artint.2007.10.009.

McCarthy, J., M. L. Minsky, N. Rochester, and C. Shannon (1955). A proposal for the Dartmouth summer research project on artificial intelligence, August 31, 1955. *AI Magazine 27*(4), 12.

McCarthy, J. and C. Shannon (1956). Preface. In C. Shannon and J. McCarthy (Eds.), *Automata Studies.* Princeton: Princeton University Press.

McDermott, D. (2007). Level-headed. *Artificial Intelligence 171*(18), 1183–1186.

McDermott, D. (2014). What was Alan Turing's imitation game? *The Critique.* Part of an issue on Turing's imitation game. Available at: http://www.thecritique.com/articles/what-was-alan-turings-imitation-game/. Access on 20 January 2021.

McEwan, I. (2019). *Machines Like Me.* London: Jonathan Cape.

McGuinness, B. (1988). *Wittgenstein: A Life: Young Ludwig 1889–1921.* Berkeley: University of California Press.

Michie, D. (1984, 9 Feb.). A loner, a misfit, a genius. *The New Scientist*, 36–37. Available at: https://books.google.co.uk/books?id=6idhDboGmZoC&pg=PA36. Access on 11 May 2023.

Michie, D. (2002). Transcript of interview. In *Recollections of early AI in Britain: 1942–1965 (video for the BCS Computer Conservation Society's October 2002 Conference on the history of AI in Britain).* BCS Computer Conservation Society. Available at: http://www.aiai.ed.ac.uk/events/ccs2002/CCS-early-british-ai-dmichie.pdf. Access on 31 May 2022.

Millar, P. H. (1973). On the point of the imitation game. *Mind LXXXII*(328), 595–597. doi: 10.1093/mind/LXXXII.328.595.

Millican, P. (2021). Alan Turing and human-like intelligence. In S. Muggleton and N. Chater (Eds.), *Human-Like Machine Intelligence*, pp. 24–51. Oxford: Oxford University Press. doi:10.1093/oso/9780198862536.003.0002.

Millican, P. and A. Clark (Eds.) (1999). *The Legacy of Alan Turing: Connectionism, Concepts and Folk Psychology, Volume 2 of Legacy of Alan Turing.* Oxford: Clarendon Press.

Minsky, M. (2013, July). Marvin Minsky on AI: The Turing test is a joke! Interview to the *Singularity Weblog.* Available at: http://www.singularityweblog.

com/marvin-minsky/. From 23'35" to 24'45". Access on 20 June 2023.

Moor, J. (Ed.) (2003). *The Turing Test: The Elusive Standard of Artificial Intelligence*. Studies in Cognitive Systems. Dordrecht: Springer.

Moor, J. H. (1976). An analysis of the Turing test. *Philosophical Studies 30*(4), 249–257. doi: 10.1007/BF00372497.

Moor, J. H. (2001). The status and future of the Turing test. *Minds and Machines 11*(1), 77–93. doi: 10.1023/A:1011218925467.

Mountbatten, L. (1946). The presidential address. *Journal of the British Institution of Radio Engineers 6*(6), 221–225. doi:10.1049/jbire.1946. 0032.

Nagel, E. and J. R. Newman (1958). *Gödel's Proof*. New York: University Press.

Naylor, R. (1976). Galileo: real experiment and didactic demonstration. *Isis 67*(3). doi:10.1086/351631.

Naylor, R. H. (1974). Galileo and the problem of free fall. *British Journal for the History of Science 7*(2), 105–134. doi: 10.1017/S0007087400013108.

Newman, M. (1955). Alan Mathison Turing, 1912–1954. *Biographical memoirs of fellows of the Royal Society 1*(November), 252–263. doi:10.1098/rsbm.1955.0019.

Newman, M. H. A. (1949, June 25). A note on electronic automatic computing machines. *British Medical Journal 1*(4616), 1133.

Newman, W. (2002). Married to a mathematician: Lyn Newman's life in letters. *The Eagle. Cambridge: St John's College 48*, 47–55. Available at: http://www.joh.cam.ac.uk/sites/default/files/Eagle/Eagle%20Volumes/2000s/ Eagle_2002.pdf. Access on 31 August 2022.

Newman, W. (2012). Alan Turing remembered: a unique firsthand account of formative experiences with Alan Turing. *Communications of the ACM 55*(12), 39–40. doi: 10.1145/2380656.2380682.

Norton, J. (1996). Are thought experiments just what you thought? *Canadian Journal of Philosophy 26*(3), 333–366. doi: 10.1080/00455091.1996.10717457.

NPL (1946, 26 Nov.). A.C.E. project – origin and early history. Technical report, National Physical Laboratory, Teddington. Public Record Office, Kew, Richmond, Surrey (document reference DSIR 10/385). Facsimile available at: www.AlanTuring.net/ace_early_history. Access on 1 December 2019.

Palmieri, P. (2005a). Galileo's construction of idealized fall in the void. *History of Science 43*(4), 343–389. doi:10.1177/007327530504300401.

Palmieri, P. (2005b). 'Spuntar lo scoglio più duro:' did Galileo ever think the most beautiful thought experiment in the history of science? *Studies in the History and Philosophy of Science Part A 36*(2), 223–240. doi:10.1016/j.shpsa.2005. 03.001.

Palmieri, P. (2018). Galileo's thought experiments: projective participation and the integration of paradoxes. In M. T. Stuart, Y. Fehige, and J. R. Brown (Eds.), *The Routledge Companion to Thought Experiments*, Chapter 5, pp. 92–110. Routledge.

Peden, K. (2011). Descartes, Spinoza, and the impasse of French philosophy: Ferdinand Alquié versus Martial Gueroult. *Modern Intellectual History 8*(2), 361–390. doi:10.1017/S1479244311000229.

Pettit, P. (1991). Realism and response-dependence. *Mind 100*(4), 587–626. https://www.jstor.org/stable/2255012.

Piccinini, G. (2000). Turing's rules for the imitation game. *Minds and Machines 10*(4), 573–582. doi: 10.1023/A:1011246220923.

Pinch, T. (2015). Scientific controversies. In J. D. Wright (Ed.), *International Encyclopedia of the Social & Behavioral Sciences* (Second ed.). Elsevier. doi:10.1016/B978-0-08-097086-8.85043-6.

Pinsky, L. (1951). Do machines think about machines thinking. *Mind 60*(239), 397–398. doi:10.1093/mind/LX.239.397.

Polanyi, M. (1974 [1958]). *Personal Knowledge: Towards a Post-Critical Philosophy* (Second ed.). Chicago: University of Chicago Press.

Popper, K. (2002 [1959]). *The Logic of Scientific Discovery*. Routledge Classics, Volume 56. London: Routledge. English edition translated and extended from the 1935 German edition *Logik der Forschung: zur Erkenntnistheorie der Modernen Naturwissenschaft*. Wien: Springer-Verlag.

Price, D. J. d. (1964). Automata and the origins of mechanism and mechanistic philosophy. *Technology and Culture 5*(1), 9–23. doi:10.2307/310 1119.

Price, J. V. (1965). *The Ironic Hume*. Austin: University of Texas Press.

Proudfoot, D. (2011). Anthropomorphism and AI: Turing's much misunderstood imitation game. *Artificial Intelligence 175*(5–6), 950–957. doi: 10.1016/j.artint.2011.01.006.

Proudfoot, D. (2013). Rethinking Turing's test. *The Journal of Philosophy 110*(7), 391–411. https://www.jstor.org/stable/43820781.

Proudfoot, D. (2015). Mocking AI panic: Turing anticipated many of today's worries about super-smart machines threatening mankind. *IEEE Spectrum*. Available at: http://spectrum.ieee.org/mocking-ai-panic.

Proudfoot, D. (2017a). The Turing test from every angle. In B. J. Copeland et al. (Eds.), *The Turing Guide*, Chapter 27, pp. 287–300. Oxford: Oxford University Press.

Proudfoot, D. (2017b). Turing's concept of intelligence. In B. J. Copeland et al. (Eds.), *The Turing Guide*, Chapter 28, pp. 301–307. Oxford: Oxford University Press. doi:10.1093/oso/9780198747826.003.0038.

Purtill, R. L. (1971). Beating the imitation game. *Mind LXXX*(318), 290–294. doi: 10.1093/mind/LXXX.318.290.

Putnam, H. (1975 [1960]). Minds and machines. In H. Putnam (Ed.), *Philosophical Papers*, pp. 362–385. Cambridge: University Press. doi:10.1017/CBO9780 511625251.020. First published in Hook, S. (Ed.) (1960). *Dimensions of Mind*, pp. 148–179. New York: New York University Press.

Rope, C. (2010, Winter 2009/10). Pioneer profiles: Douglas Hartree. *Computer Resurrection: The Bulletin of the Computer Conservation Society* (49). Available at: https://www.computerconservationsociety.org/resurrection/res49.htm. Access on 10 April 2022.

Rorty, R., J. B. Schneewind, and Q. Skinner (Eds.) (1984). *Philosophy in History: Essays on the Historiography of Philosophy*. Cambridge: Cambridge University Press.

Rowbotham, S. (2009). *Edward Carpenter: A Life of Liberty and Love*. New York: Verso.

Russell, B. (1972 [1945]). *A History of Western Philosophy*. New York: Simon & Schuster/Touchstone.

Russell, S. (2019). *Human Compatible: Artificial Intelligence and the Problem of Control*. New York: Viking.

Ryle, G. (1951). Ludwig Wittgenstein. *Analysis 12*(1), 1–9. doi:10.2307/3326710.

Ryle, G. (2000 [1949]). *The Concept of Mind*. Chicago: University Press.

Sampson, G. (1973). In defence of Turing. *Mind LXXXII*(328), 592–594.

Sargent, L. T. (2010). *Utopianism: A Very Short Introduction*. Oxford: Oxford University Press. doi:10.1093/actrade/9780199573400.001.0001.

Saunders, P. T. (1992). Introduction. In P. T. Saunders (Ed.), *Collected Works of A. M. Turing: Morphogenesis*, Volume 3, pp. xi–xii. Amsterdam: North Holland.

Saygin, A. P., I. Cicekli, and V. Akman (2000). Turing test: 50 years later. *Minds and Machines 10*(4), 463–518. doi: 10.1023/A:1011288000451.

Schurr, P. H. (1997). *So that was Life: A Biography of Sir Geoffrey Jefferson, Master of the Neurosciences and Man of Letters*. London: Royal Society of Medicine Press.

Scrivener, M. H. (2016 [1982]). *Radical Shelley: The Philosophical Anarchism and Utopian Thought of Percy Bysshe Shelley*. Princeton Legacy Library. Princeton: Princeton University Press.

Searle, J. (1980). Minds, brains, and programs. *Behavioral and Brain Sciences 3*(3), 417–424. doi: 10.1017/S0140525X00005756.

Segre, M. (1980). The role of experiment in Galileo's physics. *Archive for History of Exact Sciences 23*(3), 227–252. doi:10.1007/bf00357045.

Segre, M. (1988). Galileo as a politician. *Sudhoffs Archiv 72*(1), 69–82. doi:JSTOR 20777160.

Segre, M. (1989a). Galileo, Viviani and the Tower of Pisa. *Studies in History and Philosophy of Science Part A 20*(4), 435–451. doi: 10.1016/0039-3681(89)90018-6.

Segre, M. (1989b). Viviani's life of Galileo. *Isis 80*(2), 206–231. doi:355009 355009.

Segre, M. (1997). Light on the Galileo case? *Isis 88*(3), 484–504. doi:383771 383771.

Settle, T. B. (1961). An experiment in the history of science. *Science 133*(3445), 19–23. doi:10.1126/science.133.3445.19.

Shah, H. and K. Warwick (2015). Human or machine? *Communications of the ACM 58*(4), 8. doi:10.1145/2740243.

Shannon, C. and J. McCarthy (1956). *Automata Studies*. Princeton: University Press.

Shapin, S. (1984). Pump and circumstance: Robert Boyle's literary technology. *Social Studies of Science 14*(4), 481–520. doi:10.1177/030631284014004001.

Shapin, S. and S. Schaffer (2011 [1985]). *Leviathan and the Air Pump: Hobbes, Boyle, and the Experimental Life*. Princeton: Princeton University Press.

Shea, W. (1972). *Galileo"s Intellectual Revolution*. New York: Science History Publications.

Shelley, M. (1959 [1839]). Mrs. Shelley's note on Prometheus Unbound written for her edition of 1839. In L. J. Zillman (Ed.), *Shelley's 'Prometheus Unbound': a variorum edition*. Seattle: University of Washington Press.

Shelley, M. (2012 [1818]). *Frankenstein; or, the Modern Prometheus* (Second ed.). Norton Critical Editions. New York: W. W. Norton.

Shelley, P. B. (1820). *Prometheus Unbound, A Lyrical Drama in Four Acts with Other Poems*. London: C and J Ollier.

Shieber, S. M. (1994a, June). Lessons from a restricted Turing test. *Communications of the ACM 37*(6), 70–78. doi:10.1145/175208.175217.

Shieber, S. M. (1994b). On Loebner's lessons. *Communications of the ACM 37*(6), 83–84.

Shieber, S. M. (Ed.) (2004, June). *The Turing Test: Verbal Behavior as the Hallmark of Intelligence*. Cambridge, MA: MIT Press.

Shieber, S. M. (2007). The Turing test as interactive proof. *Nôus XLI*(4), 686–713. doi:10.1111/j.1468-0068.2007.00636.x.

Shieber, S. M. (2016). Principles for designing an AI competition, or why the Turing test fails as an inducement prize. *AI Magazine 37*(1), 91–96. doi:10.1609/aimag.v37i1.2646.

Skinner, B. F. (1957). *Verbal Behavior*. The Century Psychology Series. New York: Appleton-Century-Crofts. doi:10.1037/11256-000.

Skinner, B. F. (1991 [1938]). *The Behavior of Organisms: An Experimental Analysis*. Cambridge, MA: B. F. Skinner Foundation.

Smith, R. M. (1945). *The Shelley Legend*. New York: Charles Scribner's Sons.

Sorensen, R. A. (1992). *Thought Experiments*. Oxford: University Press.

Stalker, D. F. (1978). Why machines can't think: A reply to James Moor. *Philosophical Studies 34*(3), 317–320. doi: 10.1007/BF00372897.

Sterrett, S. G. (2000). Turing's two tests for intelligence. *Minds and Machines 10*, 541–559. doi:10.1023/A:1011242120015.

Sterrett, S. G. (2012). Bringing up Turing's child-machine. In S. B. Cooper et al. (Eds.), *How the World Computes: Proceedings of the Turing Centenary Conference, Cambridge, UK*, Number 7318 in Lecture Notes in Computer Science, pp. 703–713. Heidelberg: Springer. doi:10.1007/978-3-642-30870-3_71.

Sterrett, S. G. (2017). Turing on the integration of human and machine intelligence. In J. Floyd and A. Bokulich (Eds.), *Philosophical explorations of the legacy of Alan Turing, Volume 324 of Boston Studies in the Philosophy and History of Science*, Chapter 14, pp. 323–338. Springer. doi:10.1007/978-3-319-53280-6_14.

Sterrett, S. G. (2020). The genius of the 'original imitation game' test. *Minds and Machines 30*, 469–486. doi:10.1007/s11023-020-09543-6.

Stuart, M. T. (2018). How thought experiments increase understanding. In M. T. Stuart, Y. Fehige, and J. R. Brown (Eds.), *The Routledge Companion to Thought Experiments*, Chapter 30, pp. 526–544. London: Routledge.

Stuart, M. T. (2020). The productive anarchy of scientific imagination. *Philosophy of Science 87*, 968–978. doi:10.1086/710629.

Stuart, M. T. (2021). Telling stories in science: Feyerabend and thought experiments. *Hopos: The Journal of the International Society for the History of Philosophy of Science 11*(1), 262–281. doi:10.1086/712946.

Stuart, M. T., Y. Fehige, and J. R. Brown (Eds.) (2018). *The Routledge Companion to Thought Experiments*. London: Routledge. doi:10.4324/97813151 75027.

Swinton, J. (2019). *Alan Turing's Manchester*. Manchester: Infang Publishing.

Sykes, C. (1992, 9 March). BBC Horizon: the strange life and death of Dr. Turing. Documentary. Produced by Christopher Sykes, and edited by Jana Bennett. Available at: http://www.youtube.com/watch?v=Z-sTs2o0VuY. Access on 20 September 2022.

Symonds, J. A. (1884). *Shelley*. London: MacMillan.

Teuscher, C. (2002). *Turing's Connectionism: An Investigation of Neural Network Architectures*. Dordrecht: Springer.

Teuscher, C. (Ed.) (2006, June). *Alan Turing: Life and Legacy of a Great Thinker*. Berlin: Springer.

Toulmin, S. (1961). *Foresight and Understanding*. Bloomington: Indiana University Press.

Traiger, S. (2000). Making the right identification in the Turing test. *Minds and Machines 10*(4), 561–572. doi:10.1023/A:1011254505902.

Turing, A. (2012 [*c.* early 1952]a). Letter to Norman Routledge. In A. Hodges (Ed.), *Alan Turing: The Enigma*, pp. xxviii. Princeton: Princeton University Press. Preface to the centenary edition.

Turing, A. M. (1936). On computable numbers, with an application to the Entscheidungsproblem. *Proceedings of the London Mathematical Society s2-42*(1), 230–265. doi: 10.1112/plms/s2-42.1.230.

Turing, A. M. (1939 [1938]). Systems of logic based on ordinals. *Proceedings of the London Mathematical Society, Series 2 45*(1), 161–228. doi: 10.1112/plms/s2-45.1.161.

Turing, A. M. (1946). Letter to Ross Ashby. British Library, Collection 'W. Ross Ashby: correspondence of W. Ross Ashby', Add MS 89153/26. Facsimile available at: https://www.bl.uk/collection-items/letter-from-alan-turing-to-w-ross-ashby. Access on 1 June 2023.

Turing, A. M. (1950). Computing machinery and intelligence. *Mind 59*(236), 433–460. doi: 10.1093/mind/LIX.236.433.

Turing, A. M. (1952b). The chemical basis of morphogenesis. *Philosophical Transactions of the Royal Society B: Biological Sciences 237*(641), 37–72. doi:10.1098/rstb.1952.0012.

Turing, A. M. (2004 [1947]). Lecture on the Automatic Computing Engine. In B. J. Copeland (Ed.), *The Essential Turing: The Ideas that Gave Birth to the Computer Age*, pp. 378–394. Oxford: Oxford University Press.

Turing, A. M. (2004 [1948]). Intelligent machinery. In B. J. Copeland (Ed.), *The Essential Turing: The Ideas that Gave Birth to the Computer Age*, pp. 410–432. Oxford: University Press.

Turing, A. M. (2004 [1951]a). Can digital computers think? In B. J. Copeland (Ed.), *The Essential Turing: The Ideas that Gave Birth to the Computer Age*, pp. 482–486. Oxford: Oxford University Press.

Turing, A. M. (2004 [1953]). Chess. In B. J. Copeland (Ed.), *The essential Turing: the Ideas that Gave Birth to the Computer Age*, pp. 569–575. Oxford: Oxford University Press. Chapter 16. Facsimile available at: http://www.turingarchive.org/browse.php/B/7. Access on 20 July 2020.

Turing, A. M. (2004 [*c.* 1951]b). Intelligent machinery, a heretical theory. In B. J. Copeland (Ed.), *The Essential Turing: The Ideas that Gave Birth to the Computer Age*, pp. 472–475. Oxford: Oxford University Press.

Turing, A. M. (2005 [1945]). Proposed electronic calculator. In B. J. Copeland (Ed.), *Alan Turing's Automatic Computing Engine: The Master Codebreaker's Struggle to Build the Modern Computer*, pp. 369–454. Oxford: Oxford University Press.

Turing, A. M. et al. (2005 [1949]). Rough draft of the Discussion on the Mind and the Computing Machine, held on Thursday, 27th October, 1949, in the Philosophy Seminar. *The Rutherford Journal 1*(December). Transcript of notes taken during the philosophy seminar co-chaired by Michael Polanyi and Dorothy Emmet at the University of Manchester on 27 October 1949, made available by Wolfe Mays. Available at: http://rutherfordjournal. org/article010111.html. Facsimile available at: http://www.alanturing.net/ philosophy_seminar_oct1949/. Access on 20 July 2020.

Turing, A. M., R. Braithwaite, G. Jefferson, and M. Newman (2004 [1952]). Can automatic calculating machines be said to think? In B. J. Copeland (Ed.), *The Essential Turing: The Ideas that Gave Birth to the Computer Age*, pp. 494–506. Oxford: Oxford University Press.

Turing, S. (2012 [1959]). *Alan M. Turing: Centenary Edition*. Cambridge: Cambridge University Press.

Vardi, M. Y. (2014). Would Turing have passed the Turing Test? it's time to consider the Imitation Game as just a game. *Communications of the ACM 57*(9), 5. doi:10.1145/2643596.

Walshe, F. M. R. (1961). Geoffrey Jefferson, 1886–1961. *Biographical Memoirs of Fellows of the Royal Society 7*(November). doi: 10.1098/rsbm.1961 .0010.

Warwick, K. and H. Shah (2015). Can machines think? A report on Turing test experiments at the Royal Society. *Journal of Experimental & Theoretical Artificial Intelligence*, 989–1007. doi:10.1080/0952813X.2015.1055826.

Warwick, K. and H. Shah (2016). *Turing's Imitation Game: Conversations with the Unknown*. Cambridge: University Press.

Warwick, K., H. Shah, and J. H. Moor (2013). Some implications of a sample of practical Turing tests. *Minds and Machines 23*, 163–177.

Watson, A. G. D. (1938). Mathematics and its foundations. *Mind 47*(188), 440–451. doi:10.1093/mind/XLVII.188.440.

Weizenbaum, J. (1966). ELIZA: a computer program for the study of natural language communication between man and machine. *Communications of the ACM 9*(1), 36–45. doi: 10.1145/365153.365168.

Wheeler, M. (2020). Deceptive appearances: the Turing test, response-dependence, and intelligence as an emotional concept. *Minds and Machines 30*, 513–532. doi:10.1007/s11023-020-09533-8.

Whitby, B. (1996). The Turing test: AI's biggest blind alley? In P. Millican and A. Clark (Eds.), *Machines and Thought: The Legacy of Alan Turing*, Volume 1. Oxford: Oxford University Press.

Wiener, N. (1965 [1948]). *Cybernetics: Or Control and Communication in the Animal and the Machine* (Second ed.). Cambridge, MA: MIT Press.

Wigner, E. P. and R. A. Hodgkin (1977). Michael Polanyi, 12 march 1891–22 february 1976. *Biographical Memoirs of Fellows of the Royal Society 23*, 412–448. doi: 10.1098/rsbm.1977.0016.

Winograd, T. (1972). Understanding natural language. *Cognitive Psychology 3*(1), 1–191. doi:10.1016/0010-0285(72)90002-3. Reprinted from Winograd, T. (1972). *Understanding Natural Language*. New York: Academic Press.

Wittgenstein, L. (1958 [c. 1933–1935]). *Preliminary Studies for the "Philosophical Investigations": Generally Known as the Blue and Brown Books*. Oxford: Blackwell Publishing.

Wittgenstein, L. (2001 [1953]). *Philosophical Investigations*. Oxford: Blackwell Publishing.

Index

Note: Page numbers followed by "n" denote endnotes.

219

For Product Safety Concerns and Information please contact our EU
representative GPSR@taylorandfrancis.com
Taylor & Francis Verlag GmbH, Kaufingerstraße 24, 80331 München, Germany